OUR RESONANT UNIVERSE

This book presents the *resonance model*, a ground-breaking vision of the nature of the universe. By applying a simple electromagnetic approach to both elemental matter and nuclear reactions, Pinnow and Miller unite mass, spin, and charge with the four fundamental forces of the universe, without invoking the arbitrary constants of the standard model or the multi-dimensional mathematics of string theory. The results are striking in their accuracy and in their descriptive power. The model also provides a series of predictions that will soon put this conception of the universe to the test.

OUR RESONANT UNIVERSE

A New Look at the Unification of Physics

Douglas A. Pinnow, Ph.D.

and Kirk Miller, M.A.

GEODE BOOKS

Santa Barbara, California

OUR RESONANT UNIVERSE:
A NEW LOOK AT THE UNIFICATION OF PHYSICS

ISBN 1-4107-2634-7 (e-book)
ISBN 1-4107-2633-9 (paperback)

Library of Congress Control № 2003094633

This book is printed on acid-free paper.

Printed in the United States of America, Bloomington, Indiana

A draft of this work was published on the internet in September 1998 under the title *A Framework for a Theory of Everything: The Ultimate Quest in Physics.*

FRONT COVER: *A fanciful rendition of particle decay is traced out in the structure of galaxy NGC 1232. (European Southern Observatory)*

BACK COVER: *Galaxy NGC 3486. (Cuillandre/CFHT). Cf. figure 3-1.*

1stBooks – rev. 11/04/03

Foreword

In the first half of the twentieth century the world was presented with four fundamental forces, which provided the causal agent for all physical processes. They were gravitation, electromagnetism, and the strong and weak nuclear forces. The physics community became justifiably enamored with unifying these into a single force, or rather, in seeing them as manifestations of a single law governing the universe. The search for this law came to be referred to as the development of the unified field theory.

In the second half of the century, this quest ranked high in the priorities of the physics community. As a result, astounding progress was made in a relatively short time. The standard model, electroweak theory, *the family of* grand unified theories *(GUTs),* gauge theory, string theory, *and* M-brane theory *were all developed. Each theory provided new insights and added new bricks in the road to the unified field theory.*

However, either viewed individually or as a whole, these approaches do not bring us to the end of the road, to the unified field theory itself. This is not due to any readily apparent flaw. Rather, the approaches give an impression of incompleteness. One of them might rely too heavily on ad hoc *inputs to be the right approach. Another might ignore one of the fundamental forces because it does not fit neatly into the mathematical framework. Still another might be conceptually elegant, but mathematically so intractable that it fails to provide any testable results.*

The approach originally conceived by Douglas Pinnow, then developed and presented here by Pinnow and Miller, is another brick along the yellow brick road to the unified field theory. In my opinion, however, it is a very big brick indeed, and it represents a giant step forward. It is fundamentally different from other approaches: It is conceptually simpler. It does not invent new entities to explain the universe. It is mathematically tractable. And its predictions should be testable in the near future.

Pinnow and Miller call their approach the resonance model. This is an appropriate name, as the model describes unification within the framework of a spectrum of electromagnetic standing waves. This conception, both simple and beautiful, is able to determine the masses and charges of known particles with amazing accuracy. It combines all four forces, and explains the elusive nature of electronic charge as a 'twist' in the resonating wave, a twist easily identified with a fifth dimension proposed nearly eighty years ago to explain electromagnetism in the context of general relativity.

The resonance model sets up ready tests for verification in the near future. If it is correct, do not expect to see the anticipated zoo of particles predicted by the standard model at masses just above one thousand GeV. They do not exist. If the resonance model is correct, the magnetic moment of the Σ^o particle is negative, not positive as predicted by the standard model. Both of these tests should be within the reach of experimental physics within a decade. Researchers at CERN and Fermilab, take note.

I congratulate Pinnow and Miller. In my opinion they have a well deserved place at the Round Table of the Knights of the Unified Field Theory.

Dr Kenneth S. Schneider
Greenlawn, New York

Contents

Our Resonant Universe

Preface

Great progress has been made in our understanding of the subatomic domain over the past several decades. The fruits of this work have been organized into the so-called *standard model*, the 'gold standard' of modern particle physics. However, there is growing suspicion that this will not evolve into a final theory of nature. One cause for concern is that some twenty-one *ad hoc* parameters, including the quark masses, do not emerge from fundamental assumptions.[1] This impasse has opened the door to alternative approaches to the subatomic world. *String theory* is one attempt that has stirred considerable interest, but it too has serious unresolved issues.

Thus the theoretical underpinnings of particle physics at the beginning of the 21st century are clouded by complexity and uncertainty. Steven Weinberg, who was awarded the Nobel Prize for his outstanding contributions to the standard model, said in his recent book, *Dreams of a Final Theory*, that

> the standard model involves many features that are not dictated by fundamental principles (as we would like) but instead simply have to be taken from experiment. These apparently arbitrary features include a menu of particles, a number of constants such as ratios of masses, and even symmetries themselves. We can easily imagine that any or all of these features of the standard model might have been different.

He goes on to say,

> because the standard model leaves out gravity we now think that it is only a low-energy approximation to a really fundamental unified theory.[2]

Although Weinberg holds out hope for the growing number of theoretical physicists who have focused their efforts on string theory, the late Nobelist, Richard Feynman, was highly critical of this approach. In an interview shortly before his death in early 1988, Feynman was quoted as saying,

I noticed when I was younger, that lots of old men in the field couldn't understand new ideas very well, and resisted them with one method or another, and that they were very foolish in saying these ideas were wrong—such as Einstein not being able to take quantum mechanics. I'm an old man now, and these [string theories] are new ideas, and they look crazy to me, and they look like they're on the wrong track. Now, I know that other old men have been very foolish in saying things like this, and, therefore, I would be very foolish to say this is nonsense. I am going to be very foolish, because I do feel strongly that this is nonsense! I can't help it, even though I know the danger in such a point of view. So perhaps I could entertain future historians by saying I think all this superstring stuff is crazy and in the wrong direction.

He continued,

although people say that there are no experiments to lead us, it's not true. We have some twenty-four or more—I don't know the exact number—mysterious numbers associated with masses. Why is it that the mass of the muon compared with the electron is exactly 206 or whatever it is, why are the masses of the various particles such as quarks what they are? All these numbers, and others analogous to that—which amount to some two dozen—have **no** explanations in these string theories—absolutely none! There's not an idea at the present time, in any of the theoretical structures that I have heard of, which will give a clue as to why those masses are what they are.[3]

Against this backdrop of conflict and complexity there does appear to be a consensus that pursuing a 'theory of everything' remains a worthwhile, though challenging, endeavor; and that a successful theory will be recognized if and when it is discovered. In the words of P.C.W. Davies and Julian Brown in their book *Superstrings: A Theory of Everything?*, a truly satisfactory theory of everything

should explain why physicists observe the various elementary particles that they do, and correctly predict all of their key properties such as mass, electronic charge, magnetic moment, and so on. Second, it should faithfully describe all the interactions between particles, which means that it should account for not only the four fundamental forces of nature, but also their relative strengths. Calculations with the theory also ought to yield precisely the observed values of the various inter-particle scattering amplitudes, decay rates, branching ratios, etc. In short, the theory should account for all the measured parameters of particle physics. In addition to this, it should provide an explanation for the geometry and topology of space-time, such as the number of perceived dimensions, and offer a convincing account of how the universe came into existence.[4]

The present monograph faces the concerns raised by Feynman and Weinberg head-on. It presents a fresh perspective which embraces the confirmed results of the standard model without embracing the model itself. Although the endeavor is far from finished, and not yet ready to supercede the standard model, it does go well beyond it in calculating the masses of subatomic particles and in accounting for the nuclear and gravitational forces.

This approach is also unorthodox from the point of view of gauge or string theories. It extends electromagnetic concepts to all forces, fields, and particles. The authors hope that the reader will subject this work to serious consideration and eventual verification.

The authors thank Kenneth Schneider, Norman Moyer, and Ted Rich for their insightful reviews of the manuscript, and Dr Jean-Charles Cuillandre of the Canada-France-Hawaii Telescope for his permission to reproduce the images used in Chapter 3. Kirk Miller wishes to thank Michelle and Steve for the use of their home while he finished the manuscript. Douglas Pinnow would also like to thank his son Mike for computer and logistical support. The love and patience of his wife, Joan, were essential to the success this project and are gratefully acknowledged.

Douglas Pinnow
Laguna Hills

Kirk Miller
Santa Barbara

---

[1] D.B. Cline, 'Low-energy ways to observe high-energy phenomena.'
     Scientific American (September 1994), 40-47.

[2] S. Weinberg, *Dreams of a Final Theory*
     (Pantheon Books, New York, 1992), 191-193.

[3] P.C.W. Davies and J. Brown, *Superstrings: A Theory of Everything?*
     (Cambridge University Press, 1988), 193-210.

[4] *Ibid.*, 1-5.

Synopsis

MATHEMATICS *VS* INTUITION

Adapting new mathematical methods to old problems led to a number of significant advances in physics during the second half of the twentieth century. In fact, mathematics has often overshadowed physical insight. While it may be counterproductive to criticize the resulting successes, it is the authors' opinion that mathematical methods should not be so highly regarded as to blind us to intuitive approaches, approaches that could steer us in new directions.

The fact that certain methods are mathematically consistent in no way guarantees that they will avoid contradiction in the long term or yield valid results. For example, the spectacular success of quantum electrodynamics (QED) requires the existence of the photon as the carrier of the electromagnetic force. This approach is based on a mathematical presumption that the Lagrangian, which describes the electron's wave function, is invariant under local gauge transformations. Yet early attempts to apply this method to the strong nuclear force failed.

Nonetheless, the mathematics involved in QED was so appealing that it was subsequently applied to the weak nuclear force with considerable success, producing electroweak theory.[1] Mathematically inclined physicists then applied the method to quarks, resulting in quantum chromodynamics (QCD), and then attempted to merge electroweak theory with QCD to achieve a 'grand unification theory', or GUT, which predicts, among other things, that quarks can transform into leptons.* This would mean that a proton, p^+, could decay into a pi meson, π^0, and a positron, e^+, by the process:

$$p^+ \rightarrow \pi^0 + e^+ \qquad\qquad (1).^1$$

---

* *Leptons* are particles that are not affected by the strong nuclear force, unlike mesons and baryons, but they do obey the Pauli exclusion principle. That is, they are *fermions* with half-integral spin. The electron, muon, tau, and the neutrinos are traditionally considered leptons.

The sheer beauty of the mathematics has lent credence to this GUT. However, diligent searches in the USA, Japan, India, and Europe have all failed to detect any such decay process. Their considerable efforts have determined that the proton lifetime must be greater than 4×10^{34} years.* This is many orders of magnitude beyond than the 10^{10}-year age of the universe, dimming hopes for this most straightforward approach to unification. Something is wrong—or, at the very least, the unification was based on assumptions that were only good approximations rather than fundamentally correct principles.

Perhaps, then, the multidimensional spaces of string theory are the way to go?

Brian Greene, a prolific contributor and popularizer of string theory, summarized his guarded but optimistic views in a recent book, *The Elegant Universe*:

> Currently, string theorists are in a position analogous to an Einstein bereft of the equivalence principle. Since … 1968, the theory has been pieced together, discovery by discovery, revolution by revolution. But a central organizing principle that embraces these discoveries and all other features of the theory within one overarching and systematic framework—a framework that makes the existence of each individual ingredient absolutely inevitable—is still missing. The discovery of this principle would mark a pivotal moment in the development of string theory, as it would likely expose the theory's inner workings with unforeseen clarity. There is, of course, no guarantee that such a fundamental principle exists, but [finding this] 'principle of inevitability'—that underlying idea from which the whole theory necessarily springs forth—is of highest priority.[2]

As a counterpoint, Feynman had this to say:

> [String theory] is precise mathematically, but the mathematics is far too difficult for the individuals who are doing it, and they don't draw their conclusions with any rigor. So they just guess. — So they're unable to make a precise prediction, not through carelessness but through inability. But they continue to say it looks like a promising theory, in spite of the fact that they have to add all these guesses.[3]

---

* This was determined by several very large particle detectors, containing huge numbers of protons, that have been monitored over a number of years. The largest of these, the Super Kamiokande in Japan, holds 50 000 tons of ultra-pure water.

Perhaps a dead end, perhaps not, but a very difficult road for sure.

Might there be other approaches to such a 'theory of everything'? History suggests one possibility. Major advances during the first half of the twentieth century were inspired more by physical intuition than by mathematics. Special relativity was based on the simple concept that the speed of light is the same in all inertial reference systems, and general relativity was founded on the equivalence principle. The first clear picture of the atom emerged from Bohr's intuitive model. Surely, these theories had to accurately describe nature as well as being mathematically consistent. In the case of general relativity, the mathematics is quite complex. But it follows the intuitive concept rather than forcing it.

It is the authors' belief that a theory of everything, if one exists, is likely to be based on a few broad concepts that could be grasped by most scientists. Furthermore, there would likely be some substantial progress made in establishing a basic framework for the theory, akin to the contribution of the Bohr model to quantum mechanics, or to special relativity paving the way to general relativity. It is by no means clear that string theorists will discover this 'principle of inevitability', in Greene's words, and therefore a lack of conformity with string concepts should not discourage us from exploring alternate approaches. Once a promising framework has been established, by whatever means, there will be more than enough work for the mathematically inclined to flesh it out for rigorous evaluation.

The present proposal is unorthodox. It is not guided by the string and gauge theories that are currently in favor. Instead, it extends electromagnetic concepts to encompass all forces, fields, and particles in a way that is consistent with the experimental data. Specifically, all massive particles are described as standing-wave resonances of the electromagnetic field. Thus this *resonance model* implicitly embraces the concept, usually associated with string theory, that particles are not point-like objects. This eliminates the dilemma of the infinite energies associated with the gravitational and electromagnetic fields of singularities (point charges or masses). A resolution of this difficulty is essential to any theory broad enough to account for both electromagnetism and gravitation.

Unlike string theory, which is so complex that it as yet defies quantitative predictions of even such basic parameters as the number of dimensions in our universe, the resonance model is quite specific. It makes several quantitative predictions which should either validate or falsify it in the reasonably near future.

PREDICTED RESULTS

(1) Only one new elementary particle will be observed in the 100- to 1000-GeV mass range. Many supporters of the standard model expect to find a veritable 'zoo' of new subatomic particles in this range. With considerable confidence, they have already assigned to them such names as *selectron*, *squark*, and *photino*. In contrast, the one particle predicted by the resonance model has a mass of about 780 GeV and properties similar to the Z^0. The next such particle shouldn't occur until around 1450 GeV. In view of the progress being made in high-energy experimental physics, this issue may be resolved within a decade.

(2) A series of increasingly massive elementary particles is predicted to exist well beyond the Higgs boson, one of the most massive particles predicted by the standard model.

(3) The apparent $\frac{1}{3}e$ electronic charges identified as quarks by the standard model will be complemented by other apparent odd-integral fractional charges ($\frac{1}{5}e$, $\frac{1}{7}e$, etc.), to be found in various particles heavier than the known baryons.

(4) In addition to generating the correct magnetic moments of the proton and neutron without adjustable parameters, such as are required by the standard model, the resonance model predicts a negative value for the magnetic moment of the neutral sigma baryon, which has yet to be measured because of the short lifetime of the particle. The standard model predicts a positive value.

(5) The electronic charge of the muon is predicted to be one part in 2×10^{21} less than the charge of the electron, while the charge of the tau lepton should be less by two parts in 2×10^{21}.

(6) Neutrinos have rest masses of zero. This contradicts the common interpretation of the neutrino oscillations recently observed at the Super Kamiokande detector in Japan and elsewhere. A second interpretation, consistent with zero rest mass, is provided in Chapter 4. This predicts that the oscillation length should vary inversely with neutrino energy, whereas a non-zero rest mass would imply a direct proportionality. In view of the rapid progress being made in experimental neutrino physics, this may be one of the first predictions to be evaluated.

(7) The existence of a stable neutron-like composite particle is *tentatively* predicted, with a rest mass of approximately 9.2 GeV. This new particle may account for much of the cold dark matter that is believed to exist in the universe. Furthermore, it is predicted to have a small but measurable magnetic moment, which should aid in its detection.

OVERVIEW

Chapter 1, *Elementary Particles*, draws an association between a set of allowed standing-wave electromagnetic resonances and various elementary particles, starting with the electron. A mechanism to explain the strong nuclear force in electromagnetic terms is proposed. Quarks arise naturally, and new heavy particles are predicted.

Chapter 2, *Composite Particles*, is dedicated to the interactions among the standing-wave resonances (elementary, massive particles) which form the composite baryons. Models of such structures lead to mass calculations that are accurate to within 1% of observation for all spin-½ baryons. The magnetic moments of the proton and neutron are also easily calculated from such first principles. An estimate of the magnetic moment of the neutral sigma baryon (Σ^0) is offered as a test of this approach.

The concepts of *mass, gravitation, electronic charge,* and the *cosmological implications of the resonance model* are explored in Chapter 3. It is predicted that the electronic charge of the muon is slightly less than that of the electron, by one part in 2×10^{21}. Some small difference in charge is necessary to account for parity violation, and the magnitude is suggested by the unification of the gravitational and electromagnetic forces. Chapter 3 closes with a new candidate for cold dark matter.

Chapter 4, *The Electromagnetic Field*, introduces the concept that both the photon and the neutrino are traveling-wave modes of the electromagnetic field. This chapter has more technical detail than the others, and may be skipped during an initial review without losing the main theme.

Chapter 5 recapitulates the key concepts and results of the resonance model. The authors share the thoughts that motivated their work, as well as perspectives on such abstract concepts as the Pauli exclusion principle, the conservation of strangeness, and the nature of electronic charge. They close with a reaffirmation of the traditional scientific method, in the face of a post-modern trend to judge theoretical concepts solely on the basis of mathematical consistency.

[1] G.D. Coughlan and J.E. Dodd, *The Ideas of Particle Physics*, 2nd ed.
    (Cambridge University Press, 1991), 91-101, 171-177.

[2] B. Greene, *The Elegant Universe*
    (W.W. Norton & Co., New York, 1999), 374-380.

[3] P.C.W. Davies and J. Brown, *Superstrings: A Theory of Everything?*
    (Cambridge University Press, 1988), 193-210.

Chapter 1

ELEMENTARY PARTICLES

All massive subatomic particles are equated with standing-wave resonances of the electromagnetic field. Their masses are derived from a single parameter, the rest mass of the electron, which is the lowest-order resonance. A simple explanation of the strong nuclear force in purely electromagnetic terms is proposed.

1-I. INTRODUCTION

The central premise of this monograph is the possibility that standing-wave resonances of the electromagnetic field may exist in free space and be involved in the formation of massive particles, or matter. Hence the name *resonance model*. The resonance model fully embraces quantum mechanics, electromagnetism, and relativity, but holds quantum chromodynamics, electroweak theory, and related gauge theories to be less than fundamental, although practically useful. The objective is to show that all physical forces have a common origin, and to build a model upon this foundation that explains fractional charges without promoting them to the fundamental importance of quarks. This chapter proposes a connection between the strong nuclear force and electromagnetism. Gravitation and the weak nuclear force are considered in Chapters 3 and 4, respectively.

1-II. BASIC PRINCIPLES APPLIED TO THE ELECTRON

The resonance model extends the broadly accepted premise that the electron and positron are purely electromagnetic in origin. This is suggested by the fact that an electron-positron pair, $e^- e^+$, can be generated from an electromagnetic field, and that they can return to the field through an annihilation process which results in a pair of oppositely directed gamma rays, γ,

$$e^- + e^+ \leftrightarrow \gamma + \gamma \qquad\qquad\qquad (1\text{-}1).^{[1]}$$

Central to the resonance model is the concept that the electron and positron may form from the transitory overlap of two circularly polarized traveling waves. At optical frequencies such an overlap may occur when a circularly polarized plane wave is reflected at normal incidence from a flat metallic surface. At the higher frequencies relevant to particle formation, a gamma ray may be scattered by a massive particle for the same effect. The standing wave results from the interference of the overlapping incident and reflected portions of the traveling wave. It takes the form of a rotating and radially directed electromagnetic field, as in Figure 1-1(a).

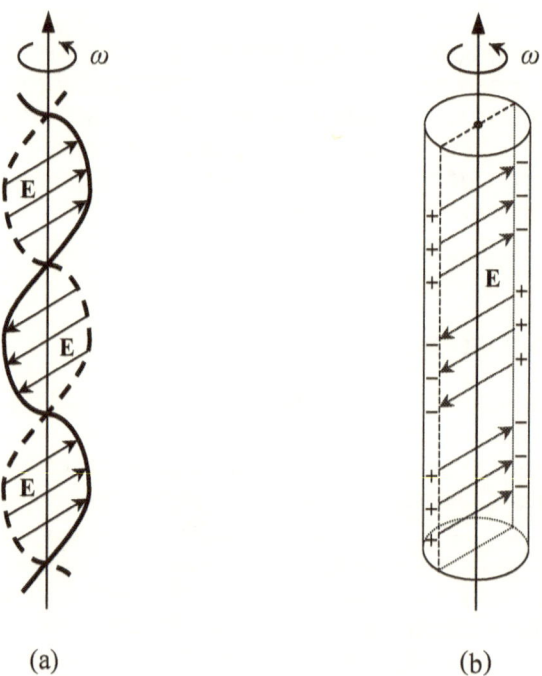

(a)   (b)

Figure 1-1. *ELECTRON-POSITRON PAIR PRODUCTION*

(a) *The electric field, **E**, along several segments of a standing wave produced by the superposition of two oppositely directed and circularly polarized electromagnetic waves. Here the upward component is right polarized and the downward component is left polarized. The resultant field rotates in place at a frequency ω.*

(b) *A Gaussian cylinder constructed with a surface rotating at the speed of light at radius r = c/ω, and with a surface electronic charge chosen to reproduce the electric field of the standing wave in (a). Note that the electric field of a charge traveling at the speed of light flattens to a plane transverse to its direction of motion.*

The participation of a massive particle is convenient for conservation of momentum.*[1] However, the standing wave is essential to the realization of an *inertial reference frame* corresponding to the center of momentum of the two interacting gamma rays. An inertial reference frame is, in turn, required to define the wavelength of the standing wave and its resonant frequency.

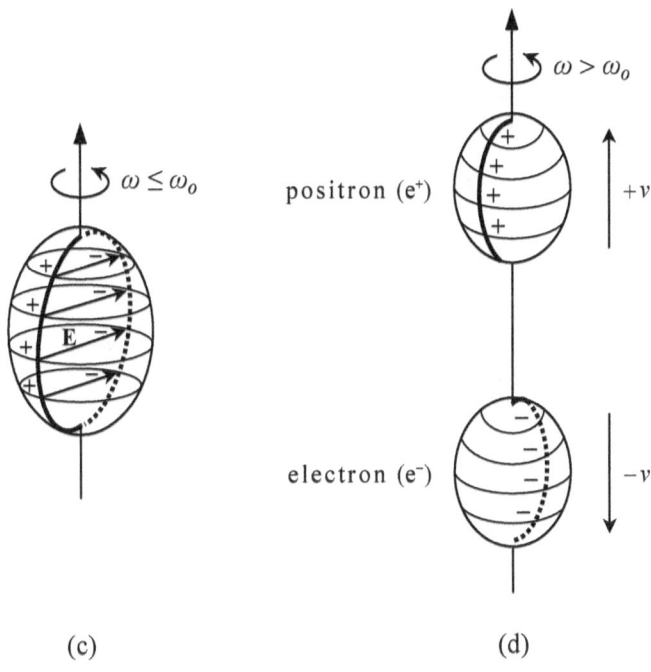

(c) (d)

Figure 1-1. (continued)

(c) *A simplified view of a half-wavelength segment from (a) and (b), showing the electronic charge confined to a two dimensional form defined by the local strength of the electric field. This shape rotates in place at a frequency $\omega \leq \omega_0$. The ends of the electric field vectors may be associated with nascent positive and negative electronic charges.*

(d) *At $\omega = \omega_0$ the velocity reaches light speed at the Compton radius, λ_c. The resonance can't spin any faster and starts to separate. At energies greater than $\hbar\omega_0$, the charges diverge with such velocities that they continue to spin at the frequency ω_0 in their respective rest frames.*

Once separated, the charges are self contained, restricted to the corporeal form of two stable particles (positron and electron), with parallel spins. See Chapter 4 for corrections to the spin equation.

---

* Some texts imply that the particle is necessary for pair production. It is merely a catalyst. Under ideal conditions, Equation (1-1) proceeds in either direction.

When averaged over time, the positive and negative 'ends' of the transverse electric field in Figure 1-1(a) can be associated with nascent positive and negative charges, superimposed in space about a common center but separated in phase by π, as shown in Figures 1-1(b-c). This is suggested by Gauss' Law, which states that electronic charge may be equated to the normal component of an electric field integrated over any imagined closed surface, such as the cylindrical surface in Figure 1-1(b).[2] For first-order resonances, a surface may be selected that is one half wavelength long and equal in circumference to the distance light travels during one rotational period, as in Figure 1-1(c).* A more complete description is given in Chapter 4, along with the origin of electronic charge.

Experimentally it is known that the threshold for particle formation occurs when the frequency of the standing-wave resonance reaches ω_0, the *natural frequency* of the electron. At this frequency the angular momentum of each particle is $\frac{1}{2}\hbar$, and the circumference of the surface in Figures 1-1 equals the Compton wavelength of the electron.[†] The frequency ω_0 is not predicted by the resonance model. Rather, it is believed to be related to the initial conditions of our universe, and will be explored further in Chapter 3.

Electrostatic attraction is insufficient to bind the charges at energies greater than $\hbar\omega_0$. Any additional energy is kinetic, such that the charges diverge at sufficiently relativistic speeds to maintain the resonant frequency ω_0 in their inertial reference frames, as in Figure 1-1(d). (In Figures 1-1, the spin of the combined electromagnetic waves is $2\hbar$, while that of the resulting particles is $\hbar$. Such spin discrepancies are resolved in Chapter 4.) The inertial frame, or *rest frame*, that is required to define ω_0 is also needed to define rest mass.

---

* Parallel electronic currents attract.[4] At a certain current magnitude—which for a rotating charge means at some rotational frequency, ω, this effective surface tension counterbalances the electrostatic self-repulsion of the charge, and the body is stable. The balance occurs when the current velocity is light speed, c:

The attractive force between two equal and parallel currents, I, separated by a distance d, is $F_B = \mu I^2/2\pi d$, and the repulsive force between two similar line charges of charge density λ is $F_E = \lambda^2/2\pi\varepsilon d$. Setting $F_B = F_E$, and given $\mu\varepsilon = 1/c^2$, we find $I = c\lambda$. That is, the surface current, I, has a velocity of c at a radius $r = c/\omega$.

† The radius $r = c/\omega$ equals the reduced Compton wavelength of the electron, $\hbar/m_e c$ = 0.39×10^{-12} m, when $\omega = \omega_0$. This is several hundred times larger than the measured radius of the proton, 1.5×10^{-15} m, yet only a fraction of a percent of the 2.5×10^{-10} m quantum-mechanical radius of the electron cloud in a hydrogen atom. That is, the electron is too big to fit inside an atomic nucleus and too small to effect the wave functions of valence electrons. Thus the resonance model leaves intact the usual interpretation of the wave function of an electron in an atom as the probability of finding the electron at some position relative to the nucleus, while ensuring that it does not react with that nucleus.

The emerging picture of the electron is a bundle of electromagnetic energy, confined to one-half a Compton wavelength along its axis of rotation, with a transverse circumference of a full Compton wavelength. The energy may be thought of as circulating within this space. This model avoids the pitfalls of early attempts to describe the electron as a rigid particle, such as the uniformly charged sphere conceived by J.J. Thompson in 1881 and further developed by Max Abraham. Abraham declared in 1902 that "the mass of the electron is of purely electromagnetic nature",[3] but shortly thereafter his model was found to behave incorrectly under Lorentz transformation (*i.e.*, relativity). This failure is common to all rigid bodies,[4] and the field has largely become dormant. Work has since shifted to Dirac's model of the electron as a point charge, despite broad concerns about its feasibility: point charges (singularities) generate infinite field strengths. Since the structure of the electron proposed by the resonance model is derived directly from the electromagnetic field, it avoids both the transformation problem of rigid bodies and the infinite field strengths that crop up with singularities.

The rest mass of the electron, m_e, is related to the resonant frequency, ω_0, as

$$m_e = \frac{\hbar \omega_0}{c^2} \tag{1-2a},$$

which can be rewritten in an equivalent but less familiar form,

$$m_e = \frac{\alpha^2 \hbar^3 \omega_0}{e^4} \tag{1-2b},$$

where α is the fine structure constant,

$$\alpha \equiv \frac{e^2}{c\hbar} \tag{1-2c}.$$

If such a stable electromagnetic resonance does exist, then higher-order resonances comprising multiple half-wavelength segments would follow.*

The theoretical considerations in Section 1-V establish that only resonances with an odd number of segments may exist. Thus, after the single-segment electron, the next resonant structure would consist of three segments that each bear ⅓ of the total electronic charge.†

---

* A *segment* is the vibrating portion between two nodes of a wave, and so by definition is one-half wavelength long.

† Electronic charge is distributed on the much smaller scale of the Planck length, and should therefore be evenly shared among the segments. (See Chapter 5.)

That is, the effectively fractional charge in each segment would be

$$e' = \tfrac{1}{3}e \qquad\qquad (1\text{-}3).$$

1-III. MU LEPTONS AND PI MESONS

Various strategies were explored to calculate the rest masses and other characteristics of higher-order resonances in terms of the known parameters of the electron. The only approach that did not seem contrived was to assume that the constants in the numerator of Equation (1-2b)—that is, α, $\hbar$, and ω_o—remain truly constant. With this assumption, the fractional charge of Equation (1-3) can be substituted into Equation (1-2b) to calculate the rest mass, m', of each segment of the resonance,

$$m' = \frac{\alpha^2 \hbar^3 \omega_o}{(e')^4} = 81 m_e \qquad\qquad (1\text{-}4).$$

Since there are three segments in this example, the *base mass* would be $3m'$, or approximately $243 m_e$ (124.2 MeV*). This base value is only a first-order approximation because it does not take into account any intersegmental attraction or repulsion.

One can tentatively associate the muon with the case where all three fractionally charged segments are polarized (circulate) in the same direction, because the muon has a spin similar to the electron. In Appendix A it is shown that the circulating electromagnetic currents in adjacent segments of the muon will cause an electromagnetic attraction in excess of electrostatic repulsion. This leads to a rest mass that is lower than the base value of three non-interacting segments.

At first there appears to be a major flaw in identifying the muon with the three-segment resonance, because the magnitude of the intersegmental electromagnetic attraction is about two orders of magnitude too small to explain the difference between the rest mass of the muon, 105.7 MeV,[5] and the base value of 124.2 MeV.

This discrepancy can be resolved by noting that if α and $\hbar$ are truly constant, then, from the definition of α in Equation (1-2c), the ratio e^2/c must remain constant even though e is permitted to vary. (The constancy of the ratio e^2/c is further supported in Section 1-V.) This leads to the striking conclusion that the speed of light within a resonance may also differ from c. If the

---

* Although it is common to speak of mass this way, it is actually the mass energy, mc^2, that is measured in electron volts. *Mass* is technically in units of eV/c^2.

electronic charge in one segment of the three-segment resonance, e', is ⅓ of the usual value e, then the group velocity within the resonance, c', becomes

$$c' = \tfrac{1}{9}c \tag{1-5a},$$

while the phase velocity, C', is

$$C' = \frac{c^2}{c'} = 9c \tag{1-5b}.[6]$$

For a resonance of n segments, this may be generalized to

$$C' = n^2 c \tag{1-6}.$$

(Note that *phase* velocities greater than c are a standard result of electromagnetic theory and do not violate the principles of relativity, since they can have no impact on causality.[7])

A group velocity $c' = \tfrac{1}{9}c$ represents the maximum free-space velocity in a hypothetical universe where the basic unit of charge is $e' = \tfrac{1}{3}e$. As such it has little practical significance. However, electronic currents and forces are dependent on the phase velocity, so this *is* significant.[8] In a three-segment resonant particle, $C' = 9c$ means that the circulating electronic current, I', is nine times greater than the expected current, I, thereby increasing the electromagnetic energy contribution, W, by a factor of $9^2 = 81$:

$$\frac{W'}{W} = \left(\frac{I'}{I}\right)^2 = \left(\frac{C'}{c}\right)^2 = \left(\frac{n^2 c}{c}\right)^2 = n^4 = 81 \tag{1-7}.$$

In Appendix A it is shown that this enhancement factor is sufficient to account for the discrepancy between the base mass of 124.2 MeV and the observed mass of the muon. A sketch of the muon based on this model is presented in Figure 1-2(a).

If, like the electron, the muon, μ, has a completely electromagnetic origin, then it follows that the two neutrinos, ν_e and $\bar{\nu}_\mu$, formed in the reaction

$$\mu^+ \rightarrow e^+ + \nu_e + \bar{\nu}_\mu \tag{1-8}[5]$$

have purely electromagnetic origins as well. Carrying this rationale a step further, the charged pion, $\pi^\pm$, which decays into a muon and a neutrino by

$$\pi^+ \rightarrow \mu^+ + \nu_\mu \tag{1-9},[7]$$

is also electromagnetic, and pion interactions should be governed by electromagnetism. This argues that the strong nuclear force, which governs the interactions between pions, is electromagnetic in nature.

According to the standard model, the π^+ is composed of two quarks, an up quark (charge $+\frac{2}{3}e$, mass ≈ 5 MeV, spin $+\frac{1}{2}\hbar$); and an antidown quark (charge $+\frac{1}{3}e$, mass ≈ 7 MeV, spin $-\frac{1}{2}\hbar$). The remaining 128 MeV or so of the mass of the pion is associated with an exchange of bosons called *gluons*, the carriers of the strong force in QCD.[9]

In the resonance model, on the other hand, a π^+ may be conceived of as a superposition of two half muons with opposite spin, $\frac{1}{2}\mu^{+\uparrow}$ plus $\frac{1}{2}\mu^{+\downarrow}$, generating the structure shown in Figure 1-2(b). Not only is the resulting net spin zero, like a pion, but it follows that there is no intersegmental electromagnetic attraction. However, the electrostatic repulsion energy would push the mass of the π^+ over the base value of 124.2 MeV. The enhancement factor of 81 from Equation (1-7) accounts well for the observed mass of the π^+. (See Chapter 2 and Appendix A for details.)

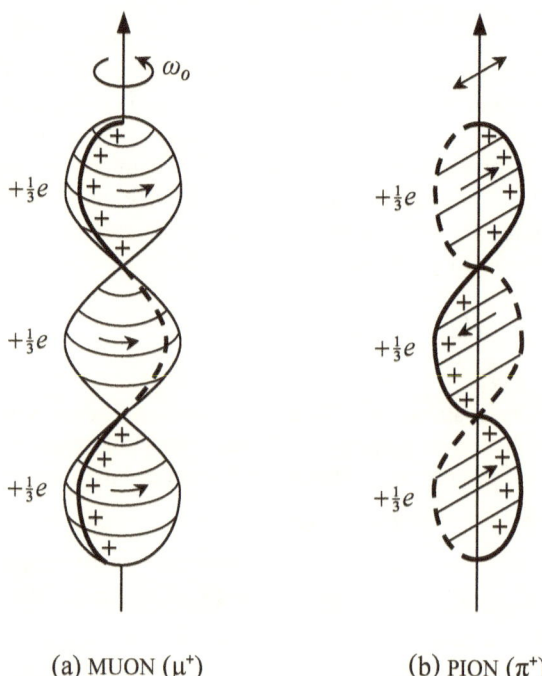

(a) MUON (μ^+)   (b) PION (π^+)

Figure 1-2. COMPARISON *of a mu lepton and a pi meson.*

(a) *Proposed structure of the μ^+ lepton. The dark line, representing an element of the charge surface, rotates about the axis. The resulting parallel loops of surface current generate intersegmental attraction.*

(b) *Proposed structure of the π^+ meson, equivalent to a superposition of two half-muons with opposite spin. The resulting surface element oscillates across the axis. Since the currents in adjacent segments are oppositely directed, they generate intersegmental repulsion.*

The two counter-rotating components of the pion produce an oscillating linear current. Thus the pion takes on a planar structure, unlike the volume swept out by the rotating muon. This mechanism may provide insight into why the Pauli exclusion principle applies to Fermi-Dirac particles *(fermions)*, like the electron and muon, which are circularly polarized and thus inherently three dimensional, but not to Bose-Einstein particles *(bosons)*, like the pion, which are basically two-dimensional and can therefore merge together at low temperatures or high pressures.

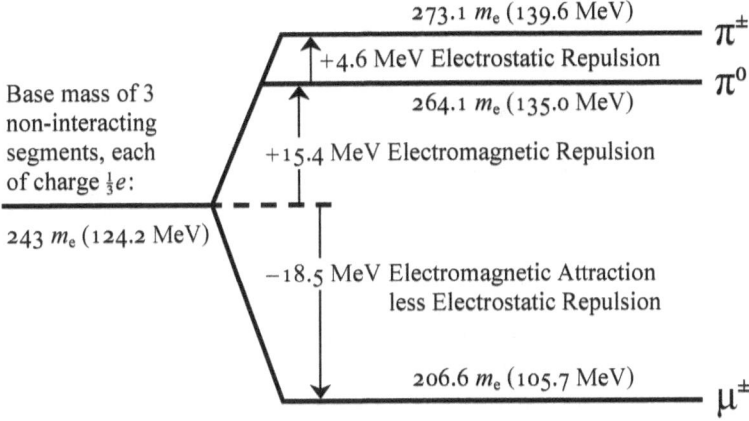

Figure 1-3. *PARTICLE MASSES of the second generation, with the base mass on the left determined by Equation (1-4), and the observed masses on the right. The masses differ from the base value because of electromagnetic attraction between segments of the circularly polarized muons, but not of the linear pions; electromagnetic repulsion between segments of the three pions; and electrostatic repulsion in the charged muons and the two charged pions.*

These standing-wave structures also offer a simple way of visualizing the nature of the strong nuclear force that acts between pions. It is simply the attraction of the parallel electronic currents in the adjacent planar pions, an attraction that is enhanced 81-fold. This is consistent with a number of survey articles indicating that the strength ratio of the strong and electro-magnetic forces is approximately two orders of magnitude.[10] Note that the muon would not experience a similar strong attraction: since the circularly polarized muon spins, interactions involve only the fraction of the current that is in the arc segments nearest the points of contact. Interactions with the linearly polarized pions, on the other hand, engage the current across their full planar surfaces. This insight is developed further in Chapter 2.

Yet another superposition, of a $\frac{1}{2}\pi^+$ and an orthogonally oriented $\frac{1}{2}\pi^-$, would have all the characteristics of the neutral π^0, including a rest mass somewhat lower than the charged π^+ due to the absence of intersegmental electrostatic repulsion,[5] and—since the charge requirements of freely traveling waves are met—a strong preference for decay into pure electromagnetic radiation. The presumed orthogonal orientation of the oppositely charged components is also consistent with the experimental observation that the two oppositely directed gamma rays emitted when a π^0 decays have linear polarizations that are orthogonally oriented.[11]

The various electromagnetic factors that cause the masses of the muons and pions to vary from the base 243-m_e value are summarized in Figure 1-3.

1-IV. HIGHER-ORDER RESONANCES

The analysis above can be easily extended to predict the base mass, m_n, of a resonance consisting of n segments. This mass would be the sum of n half-wavelength segments, each of a mass m' obtained by substituting a charge of $e' = \frac{1}{n}e$ into Equation (1-4). Thus,

$$m_n = nm' = n(n^4 m_e) = n^5 m_e \qquad (1\text{-}10).$$

The first-order approximations for the masses of the first ten fundamental resonances using this relationship are summarized in Table I. This is the full range of energies that can be explored with the high-energy physics facilities now available or under construction.

TABLE I. Base Mass Values.

Generation	Number of Segments, n	Base Mass, m_n (MeV/c^2)
1st	1	0.51
2nd	3	124
3rd	5	1 600
4th	7	8 590
5th	9	30 200
6th	11	82 300
7th	13	190 000
8th	15	388 000
9th	17	726 000
10th	19	1 270 000

Each value of *n* defines a generation. As in the case of the muon (*n* = 3), there are expected to be charged leptons with masses less than the corresponding bosons for each generation, due to attraction between the parallel circular currents in adjacent segments.* In addition, the charged bosons should have somewhat greater masses than their neutral partners, due to intersegmental electrostatic repulsion in the charged particles.

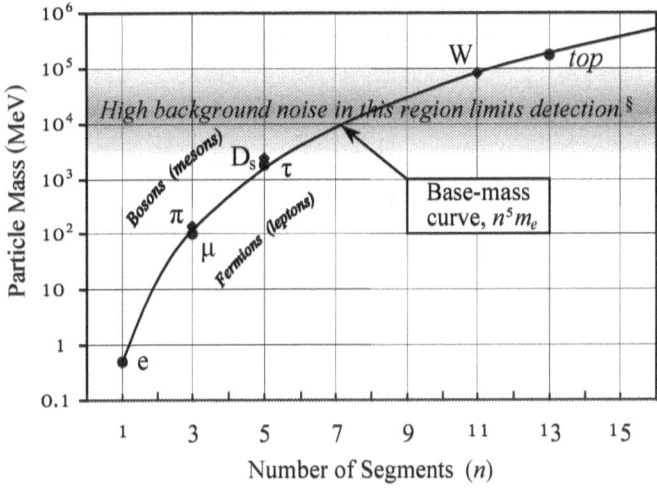

Particle	$e^\pm$	$\mu^\pm$	$\pi^\pm$	$\tau^\pm$	$D_s^\pm$	$W^\pm$	top (t)
Mass (MeV)	0.511	105.7	139.6	1777.1	1968.8	80 400	174 000

Figure 1-4. REST MASSES *of the charged particles tentatively identified as fundamental to the various particle generations. The inclusion of the D_s meson is provisional. (But see the addendum to this chapter on page 27 for an alternative.) The meson generations are expected to continue to very high energies. At present there is insufficient data to draw any firm conclusions about the number of leptons, although in the standard model they terminate after three generations.*

---

* It is possible that this binding energy limits the number of lepton generations, as the standard model maintains, if the relativistic BE enhancement were to increase at a greater rate than the $n^5 m_e$ base mass.

§ Many particles are so ephemeral that they do not leave measurable trails in even the most sophisticated detectors. They may be indirectly detected in accelerators by finding above-average numbers of decay products transverse to the colliding particle beams. While this technique works quite well at energies above about 35 MeV, the particle signature becomes increasingly obscured at lower energies where many more background events occur. While most low-mass particles last long enough to leave observable trails, there remains an intermediate range of energies, shaded on the chart, where neither method works well.

Figures 1-4 and 1-5 assign some likely correspondences to the base mass values from Table I.[12] Particle identification and placement may require adjustment as more data become available, but the fermions and bosons in Figure 1-4 are believed to be elementary resonances. They include the *top*, t, which at 174 GeV is the most massive subatomic particle identified to date.* Di-fermion resonances, such as positronium in the first generation, are plotted in Figure 1-5. The kaons and baryons are excluded from these figures because they are believed to be composites of the second-generation pions and muons, and are described in Chapter 2.

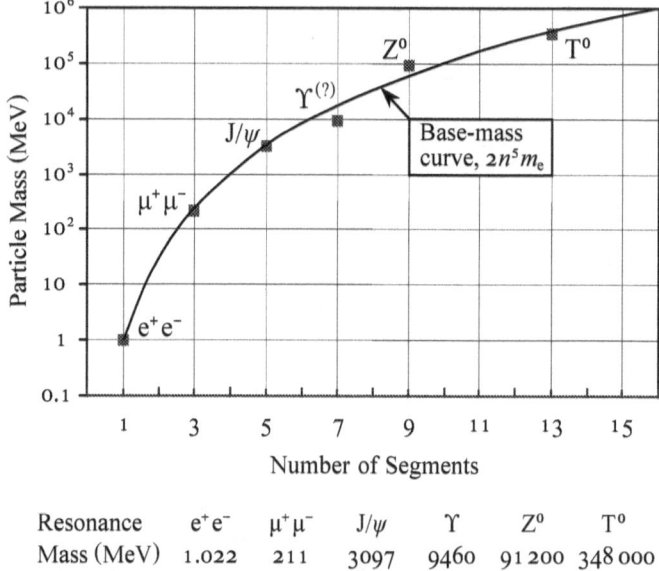

Resonance	e^+e^-	$\mu^+\mu^-$	J/ψ	Υ	Z^0	T^0
Mass (MeV)	1.022	211	3097	9460	91 200	348 000

Figure 1-5. POSITRONIUM (e^+e^-), di-muonium $(\mu^+\mu^-)$, and related di-fermion resonances which decompose into fermion-antifermion pairs. The inclusion of the Υ is provisional, as it may be a composite. The T^0 is identified as a top-antitop quark resonance $(t\bar{t})$ by the standard model.

---

* In February, 1995, two competing research teams announced the discovery of the top quark.[13] Both teams used the Fermilab Tevatron, in which 900-GeV beams of protons and antiprotons collide. At around 350 GeV, particle production events were detected in excess of the expected background noise. These were interpreted as top-antitop quark resonances, $t\bar{t}$, using hypotheses based on the standard model. However, there was no way to confirm that the participating particles had the electronic charges of $\pm\frac{2}{3}e$ or the spins of $\frac{1}{2}\hbar$ required for quarks. Because of the lack of supporting evidence, several researchers prefer to drop the 'quark' label and refer to the particles as the *top* and *antitop*. In view of the concepts developed in this section, it is possible that the top and antitop may actually be high-order leptons with unit electronic charge.

The Z^0 is one of the more difficult particles to interpret. According to the standard model, it is closely related to the $W^\pm$.[14] While this identification may seem plausible in the terms of electroweak theory, it makes little sense in the framework of the resonance model. One objective of this monograph is to develop alternatives to the unproven premises of the standard model.

The Z^0 and $W^\pm$ superficially resemble the pion family. However, the rest mass of the neutral Z^0 exceeds that of the charged Ws, whereas because of intersegmental electrostatic repulsion, it is the charged boson that would be expected to have the greater mass. The facts that the Z^0 has an angular momentum of $\hbar$, rather than 0 like the π^0, and that it decays into lepton pairs (e^+e^-, $\mu^+\mu^-$, and $\tau^+\tau^-$), also suggest that it is not a simple boson like the π^0.

Like the J/ψ and the Υ, the Z^0 has characteristics consistent with its being a high-order fermion-antifermion resonance,* similar to positronium (e^+e^-) and di-muonium ($\mu^+\mu^-$). Figure 1-5 plots these resonances along with the T^0, the designation given to the top-antitop resonance observed at 348 GeV. Although a search has never been made, it is entirely possible that the T^0 decomposes into both lepton and hadron branches as the J/ψ, Υ, and Z^0 do.

It should be pointed out that the standard model identifies the J/ψ as a charm-anticharm ($c\bar{c}$) quark resonance, the Υ as a bottom-antibottom ($b\bar{b}$) resonance, and, as just mentioned, the T^0 as a top-antitop ($t\bar{t}$) resonance. Yet there is no solid evidence to support these assignments. Individual quarks have not been observed, nor are they ever expected to be.[15]

A number of puzzles remain about the particle assignments made in Figures 1-4 and 1-5. The $D_s^\pm$ is problematic as the fundamental meson of the third generation, for example. On the plus side, it decays into a tau lepton plus a neutrino 6% of the time, paralleling the decay of the pion into a muon plus a neutrino.† However, it has no obvious neutral partner. The most obvious candidate, the D^0, is long-lived and does not decay radiatively, unlike the ephemeral π^0. The D^0 is also supposed to have an antiparticle, $\bar{D}^0$, although the evidence on this count is not clear cut.[16]

---

* These all decay into both leptons and hadrons (mesons and baryons), a behavior they share with the tau lepton. The underlying reason for this dichotomy may be that elementary bosons are equivalent to fractional leptons, $\frac{1}{2}\ell^\uparrow + \frac{1}{2}\ell^\downarrow$, as discussed in Section 1-III. In effect, boson pairs may be derivable from lepton pairs, and *vice versa*.

† There is a fundamental difference between 2nd- and 3rd-generation mesons. For the pion, no lower-generation meson decay mode is possible, since there is no meson in the first generation. Lepton decay modes are the only pathways available. The D_s, however, has multiple lower-generation pion and kaon (pion composite) pathways available to it.

(The charged D± mesons are less promising as elementary third-generation particles than the $D_s^{\pm}$ mesons, because at 1869.3 MeV they are not massive enough for tau-lepton decay modes.* Overall, the behavior of the D meson family is similar to the kaon and bottom meson families (see Section 2-VII), suggesting that they may also be composites—perhaps clusters of four kaons, as the kaons are clusters of four pions.)

Another problematic particle is the upsilon. It behaves like the other particles in Figure 1-5, but lies noticeably below the base-mass curve. An alternative interpretation of the upsilon is covered in Chapter 2.

While eventual resolution of these issues is expected, a significant point can be made at present: Many theorists who support supersymmetric extensions of the standard model expect to discover a 'zoo' of new particles in the 100- to 1000-GeV mass range. If the high-energy predictions of the resonance model have any merit, it is unlikely that many will be found. For example, the next neutral di-fermion resonance after the 348-GeV T^0 should have a mass of $2\times387 = 774$ GeV. (See Figure 1-5.) This topic is particularly timely, as physicists are just beginning to probe these energies. More results can be expected when the Large Hadron Collider comes on line within the decade at CERN.

1-V. THEORETICAL CONSIDERATIONS

G. H. Goedecke has explored solutions to Maxwell's equations that result in radiationless motion of localized charge distributions.[17] He determined very generally, by dimensional analysis, that the angular momentum of a spinning and nonradiating charge distribution must be proportional to an integer times e^2/c, independent of the size or energy content of the distribution. Goedecke's analysis does not contemplate the possibility that e and c may vary, but neither does it preclude this possibility.

One may generalize these results by hypothesizing a continuous relationship between e and c:

$$\frac{(e')^2}{c'} = \frac{e^2}{c} \tag{1-11a}$$

or $\quad c' = (e'/e)^2 c \tag{1-11b}$,

---

* According to the resonance model, a neutrino formed from D± decay would share the wavelength of the D±. Its energy would be half the mass energy of a single D± segment, or 10% of the five-segment, 1869-MeV D±. There would not be enough left over to form a 1777-MeV tau, and no tau decay modes have been observed.

while the phase velocity becomes

$$C' = (e/e')^2 c \tag{1-12},$$

where the charge ratio (e/e') can assume any value. This relationship is equivalent to Equation (1-6).

Now $C' = c$ corresponds to electromagnetic propagation in free space. Equation (1-11a) then implies that $e' = e$ for any resonances which couples to free-space electromagnetic waves. The phase-matched coupling between a propagating free-space wave and a one-segment standing-wave resonance occurs because the round-trip phase delay within the standing wave is 2π, and the incident and internal fields remain in phase at all times.

Such phase matching would occur for all standing-wave structures with odd numbers of segments. In a three-segmented resonance, for example, $C' = 9c$ and the internal field would cycle by $9 \times 2\pi = 18\pi$ during the time it would take a free-space wave to advance one wavelength (2π). Since a phase shift of 18π corresponds to three circulations within a ½-wavelength resonance, the internal and incident fields would constructively interfere, and energy from the free-space field would be able to build up within the resonance.*

In the case of resonant structures with even numbers of segments, however, no such energy buildup could occur. For a resonance two segments long, $e/e' = 2$ and C' would be $4c$. As the incident free-space field advances by π, the internal field with its phase shift of 4π would circulate once around the resonance, and the incident and internal fields would be completely out of phase. As the incident wave finishes a cycle of 2π, the internal phase shift of 8π would bring the resonance back into phase. The net result of these alternating interferences would be zero, precluding any energy buildup within the resonance. Thus only resonances with odd numbers of segments (*i.e.*, that are ½, ³⁄₂, ⁵⁄₂, ⁷⁄₂, *etc.* wavelengths in length) should exist.

Goedecke's solutions assume charge distributions with spin. This analysis is therefore applicable to electrons and muons but not to the spinless pions.

The relative stability of pions can be argued to be a consequence of their underlying equivalence to leptons, as discussed in Section 1-III, but also because of their phase-velocity mismatch (impedance mismatch) with the free-space field. A generalization of the latter concept is that spin-zero

---

* After the first internal circulation, the relative phase between the incident and circulating fields would be a partially constructive $\frac{2}{3}\pi$. After the second, it would be a partially destructive $\frac{4}{3}\pi$, negating the effect of the first. But a third circulation would bring the fields fully in phase at 2π, for a net constructive interference.

resonances could exist for all odd numbers of segments except *one*, where $C' = c$ and there is a natural impedance match. This fits with the observation that there is no meson with a rest mass similar to the electron's.*

1-VI. CONCLUSION

The resonance model describes elementary massive particles as standing-wave resonances of the electromagnetic field. The first generation of particles, consisting of a single half-wavelength segment, is associated with the electron/positron. Section 1-V argued that there can be no meson in this first generation. However, the next several generations of particles should include both mesons and leptons. In higher generations, it is possible that only mesons are to be found.

The second generation of particles can be associated with the muon and pion. The forces between the pions should be equivalent to the strong nuclear force, due to their planar structure, smaller sizes (see Appendix A) and presumably enhanced electronic currents when compared to the electron. The quark structure attributed to the pion by the standard model is also accounted for by the resonance model, where the pion is described as three contiguous segments, each with an electronic charge of $\frac{1}{3}e$. Due to intersegmental forces, the middle segment should be distinct from the two end segments. Thus in both the standard and resonance models the pion appears to contain two units: one with a $\frac{1}{3}e$ charge (the middle segment, or the down quark), and a second with a $\frac{2}{3}e$ charge (the pair of end segments, or the up quark).

The resonance model further predicts that the particles of third generation should consist of five segments, each with a charge of $\frac{1}{5}e$. The forces between these particles would not only be strong but *very strong* when compared to the muon and pion. Increasingly high-order particles could exist so long as their masses don't exceed that of the universe. This establishes an upper limit on the number of generations of elementary particles, and suggests that the decomposition of a single primordial resonance could have formed our universe. This idea will be explored in Chapter 3.

---

* *Mesons* are those bosons which are affected by the strong nuclear force. They do not obey the Pauli exclusion principle. In the resonance model, a few spin-0 mesons such as the pion are thought to be elementary particles. All spin-1 mesons, such as those included in Figure 1-5, are expected to be composites.

ADDENDUM TO CHAPTER 1

On 11 April, 2003, a paper by more than 500 co-authors from 76 different institutions was submitted to the Physical Review Letters. This paper, *Observation of a Narrow Meson Decaying to $D_s^+\pi^0$ at a Mass of 2.32 GeV/c²,* [18] is important because it describes a newly discovered particle, designated the $D^*_{sJ}(2317)^+$, that may not fit into the standard model. The most likely assignment within the Standard Model would be an excited charm-antistrange meson, $c\bar{s}$. However, after careful consideration, the co-authors conclude,

> Since a $c\bar{s}$ meson of this mass contradicts current models of charm meson spectroscopy, either these models need modification or the observed state is of a different type altogether, such as a four-quark state.

Bearing this in mind, recall the concern expressed in Section 1-IV about the identity of the fundamental second-generation boson. At the time this monograph was drafted there were no satisfactory alternatives to the 1969-MeV D_s^*. The newly discovered particle is a more promising candidate in this regard. It is interesting to note that the mass ratio of the $D^*_{sJ}(2317)^+$ to the tau lepton, 2317 MeV/1777 MeV = 1.30, would also be quite close to its first-generation equivalent, the pion-to-muon mass ratio of 1.32. Although this is suggestive, it would be premature to draw any conclusions.

[1] R.B. Leighton, *Principles of Modern Physics* (McGraw-Hill, New York, 1959), 624-626.

[2] P. Lorrain and D. Corson, *Electromagnetic Fields and Waves*, 2nd ed. (W.H. Freeman & Co., New York, 1962), 47-51.

[3] Max Jammer, *Concepts of Mass in Contemporary Physics and Philosophy* (Princeton University Press, 2000).

[4] *Ibid.*, 297-298.

[5] G.D. Coughlan and J.E. Dodd, *The Ideas of Particle Physics*, 2nd ed. (Cambridge University Press, 1991), 74-76.

[6] A. D'Abro, *The Rise of the New Physics* (Dover Publications, New York, 1951), II, 604-616.

[7] Nick Herbert, *Faster Than Light: Superluminal Loopholes in Physics* (Dutton/Plume, 1989).

[8] Lorrain and Corson, 298-299, 315-316, 356, 362.

[9] L.M. Lederman and D.N. Schramm, *From Quarks to the Cosmos* (Scientific American Library, New York, 1989), 100-113.

[10] Lederman and Schramm, 85-127.

[11] P. Hartman, *Nuclear and High Energy Particle Physics* (Cornell University, Ithaca, NY, 1959-60), 72.

[12] Coughlan and Dodd, 233-240.

[13] F. Abe *et al.*, *Physical Review Letters*, **74** (1995), 2626;

S. Abachi *et al.*, *Physical Review Letters*, **74** (1995), 2632.

[14] Coughlan and Dodd, 98-111.

[15] Coughlan and Dodd, 143-146.

[16] K. Hagiwara *et al.*, 'The Review of Particle Physics.' *Physical Review*, **D66** (2002). Available online at *http://pdg.lbl.gov*.

[17] G.H. Goedecke, 'Classically radiationless motions and possible implications for quantum theory.' *Physical Review*, **135**, 1B (13 July 1964), 281-288.

[18] The *BABAR* Collaboration, *Physical Review Letters*, **90** (2003), 242001.

Chapter 2

COMPOSITE PARTICLES

Elementary particles have been described as simple standing-wave resonances of the electromagnetic field. There are also composite particles. The baryons and the K mesons correspond to clusters of second-generation elementary particles, while the B and upsilon mesons may be composites of the third generation. This representation allows the straightforward calculation, from first principles, of both masses and magnetic moments for these particles that agree well with their measured values. The magnetic moment of the neutral sigma baryon is presented as a test of the resonance model, as this value differs substantially from the one proposed by the standard model.

2-I. INTRODUCTION

In Chapter 1, a series of subatomic particles are characterized as fundamental. Both a fermion and a boson branch are identified, each with multiple generations. The first generation consists only of the electron, while the second comprises the muon and the pion.

In this chapter the K mesons (kaons) and the baryons* are described as combinations of muons and pions. Various structures have been evaluated for fit and stability, with a bias toward efficiently packed clusters that minimize electrostatic repulsion and maximize electromagnetic attraction.

---

* *Baryons* are like mesons in that they are affected by the strong nuclear force (that is, they are *hadrons*), but on the other hand they are also *fermions* (they have half-integral spin) like the leptons. All baryons in the resonance model are composite particles, with both lepton and meson components.

This approach reaches substantially further than the standard model in that it accurately accounts for the masses and magnetic moments of these particles. In contrast, the standard model's descriptions of the kaons and baryons, developed independently by Murray Gell-Mann and Yval Ne'eman in 1961,[1] are limited to group-theory representations which are simply incapable of addressing mass or magnetic moment.

Although there are substantial similarities between the descriptions of the resonance and standard models, the differences are striking. Most significantly, the standard representation associates these particles with various multiplets of a 'special unitary group' of three abstract dimensions, $SU(3)$, while the resonance model fits within the framework of the three normal spatial dimensions.

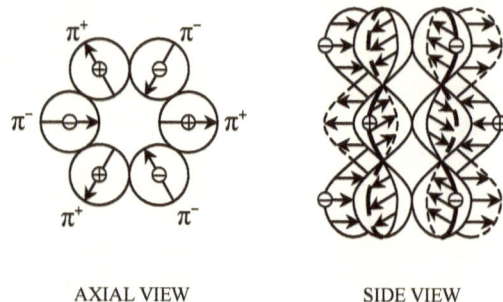

AXIAL VIEW   SIDE VIEW

Figure 2-1. RING-SHAPED CLUSTER *of six pions with alternating charges.*
The pions, depicted in more detail in Figure 1-2, attract one another electrostatically, and also electromagnetically through parallel electronic currents in adjacent pions. (The currents of inward-flowing charge in the negative pions, seen in the axial view, have vector components parallel to the outward-flowing (180° phase-shifted) charges in the positive pions.) The ring structure itself is not stable, but it is believed to constitute the outer regions of the nucleons, as discussed in Section 2-III. It may also have an ephemeral existence as the rho meson, ρ^0. Since the ρ^0 has a mass of 769 MeV, compared to a total of 837.4 MeV for six charged pions, the binding energy of a ring would be 11.4 MeV per pion.

Qualitatively, one can appreciate that the muon resonance, depicted in Figure 1-2(a), has a significantly lower rest mass than the pion because its internal electronic currents, circulating in parallel and in phase in all three segments, generate an intersegmental electromagnetic attraction. In consequence, the muon tends toward self-containment. The contrary is true for the pion, where electronic currents are oppositely directed (out of phase) in adjacent segments, producing an intersegmental repulsion, as can be seen in Figure 1-2(b). Indeed, the muon lifetime is 2.2×10^{-6} seconds, a hundred times longer than the charged pions at 2.6×10^{-8} sec. However, in properly oriented neighboring pions, these same currents could flow in phase (in the same direction) and cause strong attractive forces. This would favor planar clusters of pions side by side. Ring structures with alternatingly charged pions, as in Figure 2-1, are expected to have favorable binding energies due to both electrostatic and electromagnetic attraction. Since the positive and negative pions are not identical particles, the rings cannot collapse into boson condensates. That is, they must occupy a finite volume. The kaons can be identified with squares of four pions, the nucleons with hexagonal rings, and the sigma and xi baryons with octagonal rings.

A reality check is called for at this point. The proton is the best characterized baryon, due to its stability. Its radius, as determined by scattering experiments, is 1.2×10^{-15} m.[2] The muon and pion radii are calculated to be approximately 0.5×10^{-15} m in Appendix A, Equation (A-5), so a hexagonal ring of pions should be close to the size of the proton. Thus both the proton's mass and its size, as established through electromagnetic considerations, concur with laboratory measurement.

2-II. K MESONS (KAONS)

The structural assignments in this chapter are the result of an iterative effort to match composite pion-muon structures with the characteristics of known subatomic particles. Binding energy contributions of 15 MeV per charged pion and 12 MeV per neutral pion fit well with the observed masses of the kaons and baryons, and are in line with the theoretical expectations outlined in Appendix A. (See Equations (A-13) and (A-14) with footnote).

Structures are proposed for the four kaons in Figure 2-2. The masses are calculated by summing the individual components and subtracting the pion binding energies. For example, the K^+ is proposed to be a cluster of two π^+s, one π^-, and one π^0, for a calculated mass of

$$m_{K^+} = 2m_{\pi^+} + m_{\pi^-} + m_{\pi^0} - 3BE_{\pi^+} - BE_{\pi^0} = 496.7 \text{ MeV} \qquad (2\text{-}1).$$

This is off by only 0.6% from the observed mass of 493.7 MeV. Similarly, the K^- is composed of one π^+, two π^-s, and one π^0, with both the calculated and measured masses of the K^+.

There are two neutral kaons. The lifetime of one, called the K^0_{LONG}, is longer at 5.18×10^{-8} seconds than the other, the 0.89×10^{-10}-second K^0_{SHORT}.[3, 4] A cluster of $2\pi^+ + 2\pi^-$ is suggested for the K^0_{LONG} because its lifetime is similar to the charged pions', 2.6×10^{-8} seconds. The mass calculated for this structure is $4m_{\pi^+} - 4BE_{\pi^+} = 498.3$ MeV. This differs from the observed mass by only 0.07%. The K^0_{SHORT} is suggested to be a combination of π^+, π^-, $2\pi^0$, because π^0s in isolation live a brief 0.84×10^{-16} second. The mass calculated for the K^0_{SHORT} is 496.7 MeV, versus an observed 497.4 MeV. Again, the 0.5% variance is small. Table I summarizes the calculated *vs* observed masses of the four kaons.

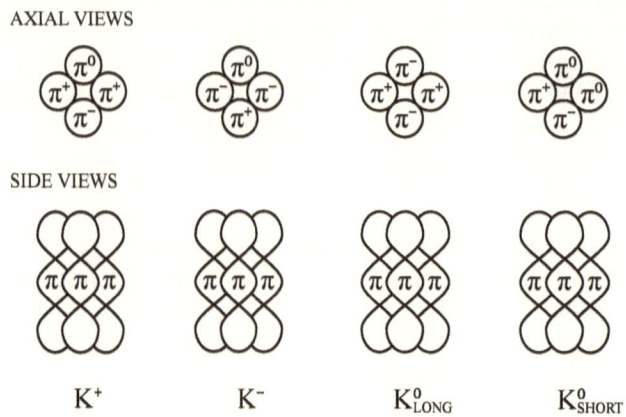

Figure 2-2. *THE STRUCTURES PROPOSED FOR THE KAONS.*

The π^0s are adjacent in the K^0_{SHORT} because it decays into gamma radiation plus either a $\pi^\pm$ pair (68% of decays) or a π^0 pair (31% of decays), and annihilation presumably occurs between adjacent components.

TABLE I. *Kaon Compositions and Masses.*

Particle	Proposed Composition	Calculated Mass (MeV)	Observed Mass (MeV)	Discrepancy
K^+	$2\pi^+ + \pi^- + \pi^0$	496.7	493.7	+0.6%
K^-	$\pi^+ + 2\pi^- + \pi^0$	496.7	493.7	+0.6%
K^0_{LONG}	$2\pi^+ + 2\pi^-$	498.3	497.9	+0.07%
K^0_{SHORT}	$\pi^+ + \pi^- + 2\pi^0$	495.1	497.4	−0.5%

This description seems simple, almost obvious. In contrast, the standard model is considerably more complex and abstract. Its $SU(3)$ representation of the kaons posits them to be members of an octet of spin-0 particles, shown in Figure 2-3(a), with each particle being located by two abstract coordinates, *strangeness* and *isotopic spin* (or *isospin*). Like all mesons, the pions and kaons are assumed to be composed of pairs or summations of pairs of fractionally charged quarks, as summarized in Figure 2-3(b).[1]

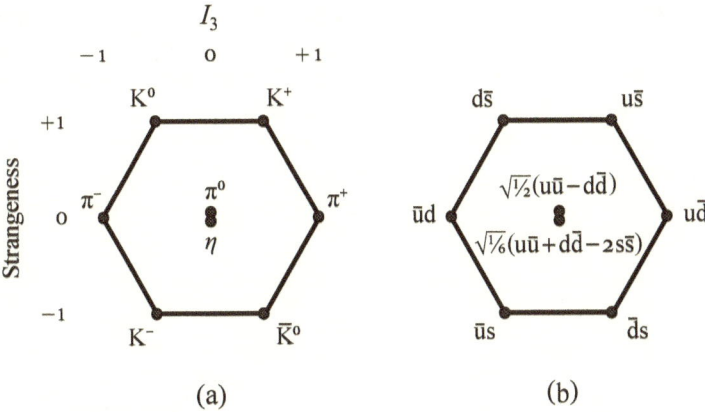

Figure 2-3. *(a) THE SPIN-0 MESON OCTET. (b) The octet's traditional quark assignments. See the endnotes for the correlation between K^0_{LONG} / K^0_{SHORT} and the $K^0 / \overline{K}^0$ depicted here.* *

Table II summarizes the quantum numbers originally assigned to the quarks of the meson octet.[1] Subsequent experimental work has shown that the spin-½ values are incorrect. This finding, which casts uncertainty on the quark model as well as the use of the $SU(3)$ representation, is referred to in the literature as the *spin crisis*.[5,6]

TABLE II. *Quark Properties.*

Quark	(q)	J	Q	I	I_3	S	B
Up	u	½	+⅔	½	+½	0	⅓
Down	d	½	−⅓	½	−½	0	⅓
Strange	s	½	−⅓	0	0	−1	⅓

J : Spin   Q : Charge
I : Isospin   S : Strangeness
I_3 : 3rd Component of Isospin   B : Baryon Number

---

* The eighth member of the octet is the electrically neutral eta meson, η. The rest mass of 547.3 MeV suggests that this particle may consist of five pions, most likely in the form of a K^0_{SHORT} with a π^0 added to the center. The central π^0 would bind strongly to the four surrounding pions, and an accurate mass cannot be calculated with the simple binding energy approximations used here. It is therefore not surprising that the estimated 621 MeV is substantially greater than the observed mass of the η.

Clearly, the resonance model's description of the spin-0 mesons is much simpler. If its scope were limited to these particles, then Occam's Razor—the principle that the simplest model to explain all relevant features is to be preferred—would favor the resonance model. The historical reason for the standard model's broad acceptance is that it characterizes not only the spin-0 meson octet, but also an octet of spin-½ baryons, including the proton and the neutron, and a decuplet of spin-³⁄₂ baryons. These are shown in Figures 2-4 and 2-5.

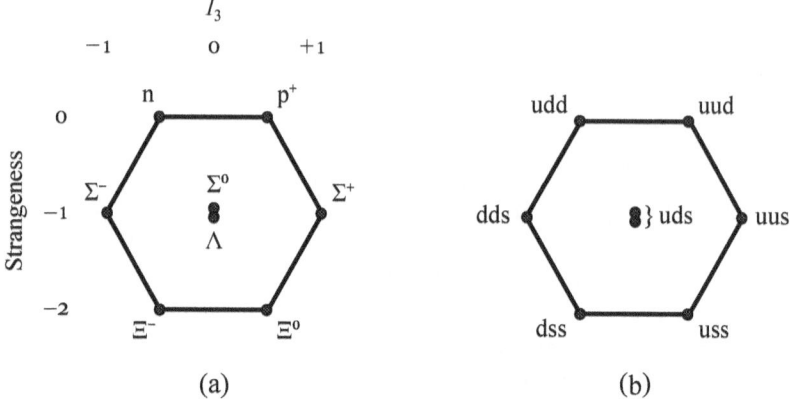

(a)   (b)

Figure 2-4. *(a) THE SPIN-½ BARYON OCTET. (b) The octet's traditional quark assignments. The proton, for example, is described as two up quarks, u, each with electronic charge +⅔, plus one down quark, d, with charge −⅓, for a net charge of +1.*

Various alternatives to the quark hypothesis of the baryons were examined, along the lines of the simple approach used to account for the meson octet.* Each baryon required a negative muon in its structure to provide the correct spin and magnetic moment.† After an iterative effort, it was found that the 15-MeV and 12-MeV binding-energy factors used for the mesons resulted in a reasonably good fit for the spin-½ baryons. However, the observed masses were consistently about 164 MeV less than expected, with the exception of the Λ, whose mass was off by −246 MeV.

---

* A recent book offers balanced views on Gell-Mann's contributions, including skeptics' positions on quarks:

But when the experiments require so many layers of interpretation, how could the physicists know when they were reading too much into the lines and squiggles [of the particle trails], seeing what their brains were wired to see, like pictures in the clouds? Were these really discoveries, or inventions?[7]

† The fact that a *negative* muon is found in the nucleons is significant to the preferential formation of matter over anti-matter in our universe. (See Chapter 3.)

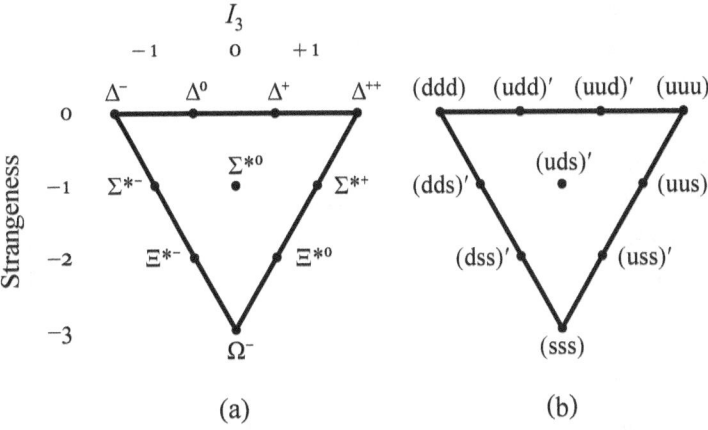

Figure 2-5. *(a)* THE SPIN-³⁄₂ BARYON DECUPLET. *(b) The traditional quark assignments. The terminology (qqq)' signifies a summation over the cyclic permutations of the quarks.*

This suggested that the muon common to the baryons might bind with its neighboring pions in such a way as to release a substantial amount of mass energy. Indeed, it turns out that the mutual annihilation (destructive interference) of one of the three segments of the muon with a segment of a neighboring pion, resulting in the loss of all mass associated with these two segments, would produce half the required energy: $\frac{1}{3}m_\pi + \frac{1}{3}m_\mu = 82$ MeV. This energy could only be lost through collisional transfer, as discussed in Appendix B. Two such segmental annihilations in the proton and neutron would account for the observed masses very well. Similar scenarios would hold for all spin-½ baryons except the Λ, where three such interferences are needed to account for the mass discrepancy: 3×82 MeV = 246 MeV.

This segmental-annihilation mechanism is analogous to pair annihilation, whereby a particle and its antiparticle—such as a π^+ and π^-, or a μ^+ and μ^-—convert completely into energy in the form of oppositely directed gamma rays. Such total mutual annihilation is not possible in the case of a π^+ with a μ^-, partly because there is no way to conserve the angular momentum of the spin-½ muon. Nevertheless, *partial* annihilation may proceed so long as the remaining segments of the π^+ and μ^- maintain the original angular momentum. This requires two surviving segments in each particle.* This mechanism is explored in Appendix B.

---

* Otherwise the surviving segments would have greater angular momentum than the $\frac{1}{2}\hbar$ spin of the muon, which was shown to be the relativistic limit in Section 1-II.

Remarkably, the rest masses calculated for the spin-½ baryons all fall to within 1% of their measured values when the binding-energy contributions in Table III are taken into account.

TABLE III. *Binding-Energy Factors in the Resonances.*

Factor	BE Contribution
$\pi^{\pm}$	15 MeV
π^0	12 MeV
$\pi^{\pm} \cdot \mu^{\pm}$ segmental annihilation	82 MeV
$\pi^0 \cdot \mu^{\pm}$ segmental annihilation	80 MeV

(80 and 82 MeV are the masses of the overlapping pion and muon segments, $(\frac{1}{3})m_{\pi} + (\frac{1}{3})m_{\mu}$. There are two annihilations in all spin-½ baryons except the Λ, which has three.)

2-III. THE PROTON AND THE NEUTRON

The structure proposed for the proton includes a central cluster of two π^+s and one μ^-, as shown in Figure 2-6, surrounded by a hexagonal ring of alternately charged pions similar to Figure 2-1. The muon provides the necessary spin. The best mass fit occurs when two segments of the muon are lost through annihilation with one segment each from the two adjacent pions. Thus the charge distribution in the central cluster appears as two units of $+\frac{2}{3}e$ and one unit of $-\frac{1}{3}e$. This distribution is experimentally observed and has been interpreted as a trio of fractionally charged quarks, a cornerstone of the standard model.[1,8] The resonance model does not require that these effectively fractional charges be elevated to the status of elementary particles.

Since collisions can carry away the mass of the lost meson segments in the form of kinetic energy, but cannot remove angular momentum, the spin of the proton must equal that of the central muon. The spin of the two lost muon segments is presumably vested within the surviving segments of the two central pions, since the remaining muon segment is fully rotating and cannot acquire additional angular momentum.

The proposed structure of the proton has a gross mass of 1222.2 MeV: two pions and one muon in the core, plus the hexagonal pion ring. Since the rest mass of the proton is 938.3 MeV, the binding energy of the structure must be 283.9 MeV. This binding energy can be accounted for by using the contributions in Table III: 15-MeV for each of the eight pions, for 120 MeV; plus the amount lost to destructive interference, $(\frac{2}{3})m_\pi + (\frac{2}{3})m_\mu = 164$ MeV. Since the muon does not participate in the strong force,* its only binding energy is presumably due to this interference. The calculated binding energy is thus 284 MeV, for a refined proton mass of 938.2 MeV— off by less than 0.02% from the observed value.

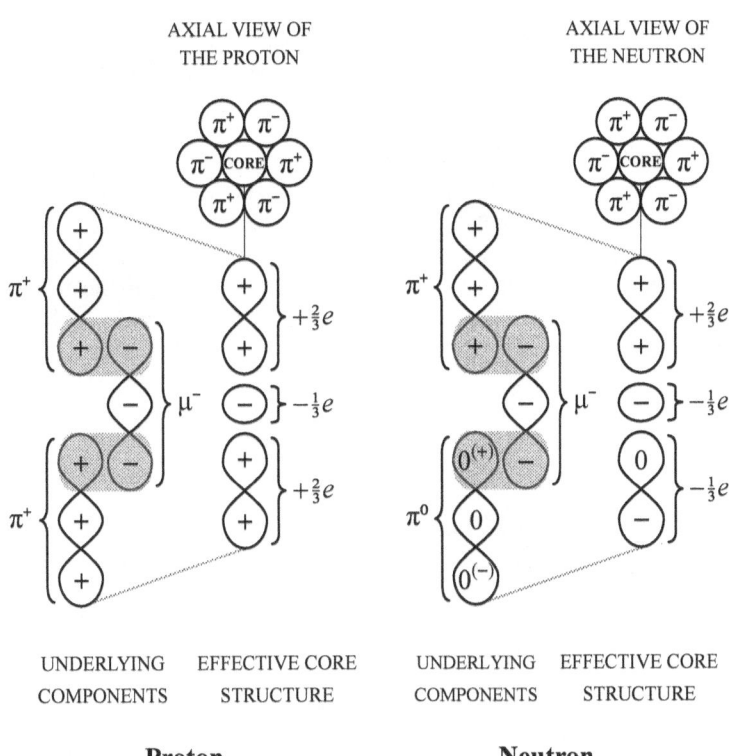

AXIAL VIEW OF   AXIAL VIEW OF
THE PROTON   THE NEUTRON

| UNDERLYING | EFFECTIVE CORE | UNDERLYING | EFFECTIVE CORE |
| COMPONENTS | STRUCTURE | COMPONENTS | STRUCTURE |

Proton   **Neutron**

Figure 2-6. *THE COMPOSITE STRUCTURES PROPOSED FOR THE NUCLEONS.*

The central clusters of muons and pions which make up the cores of the nucleons explain their quark-like behaviors. Fractional charges result from the destructive interferences, or annihilations, that eliminate the shaded segments. As this illustration suggests, the neutron has an antiparticle. That is, the antineutron, $\bar{n}$, is distinct from the neutron.

---

* The circularly polarized muon, like other leptons, has no planar surfaces and thus can make only point contact with other particles. Thus it is not subject to the augmented electromagnetic forces involved in the close planar contact between pions, and it is the pion which is the signature particle of the strong force.

The great stability of the proton can be attributed to the fact that none of its components can escape as free pions, muons, kaons, electrons, or electromagnetic radiation without an appreciable input of energy. When a π^+ decays into a μ^+ through the emission of a neutrino, 33.9 MeV of energy is released.[4] This energetically favorable reaction explains the rather short 2.5×10^{-8}-second lifetime of an unbound pion.[4] However, if this same pion were bound in the outer ring of a proton, the 33.9 MeV released would be insufficient to overcome the 284.3-MeV binding energy of the proton. Thus pion decay would no longer be energetically favorable. This proton structure would even be stable against the spontaneous mutual annihilation of neighboring oppositely charged pions in the outer ring, which would liberate $2m_{\pi^\pm} = 279.14$ MeV.

The neutron is thought to have a very similar structure to the proton. The proposed structure has a π^+, a μ^-, and a π^0 in the central cluster, as shown in Figure 2-6. This combination was selected to be consistent with the experimental finding that the neutron appears to consist of one point charge of $+\frac{2}{3}e$ and two point charges of $-\frac{1}{3}e$.[2] The mass sum of the individual components is 1217.6 MeV; the binding energy, from Table III, is 278.9 MeV;* and thus the calculated mass is 938.7 MeV—within 0.09% of the observed mass of 939.6 MeV.

These results are summarized in Table IV.

TABLE IV. *Nucleon Compositions and Masses.*
The bullets (•) represent segmental annihilation between particles.

Particle	Proposed Composition	Calculated Mass (MeV)	Observed Mass (MeV)	Discrepancy
Proton	$6\pi^\pm + [\pi^+ \cdot \mu^- \cdot \pi^+]$	938.2	938.3	-0.02%
Neutron	$6\pi^\pm + [\pi^0 \cdot \mu^- \cdot \pi^+]$	938.7	939.6	-0.09%

The electronic charge within the π^0 is assumed to redistribute itself so that $+\frac{1}{3}e$ is available in one end segment to annihilate a segment of the μ^-. In order to conserve spin and charge, either one or both of the remaining π^0 segments must then take on the spin and $-\frac{1}{3}e$ charge of the annihilated muon segment. An analysis of the magnetic moment of the neutron in Appendix C considers both possibilities.

---

* Using a BE factor of $(\frac{1}{3})m_{\pi^0} + (\frac{1}{3})m_\mu = 80.2$ MeV for the π^0 segmental annihilation, rather than the $(\frac{1}{3})m_{\pi^\pm} + (\frac{1}{3})m_\mu = 81.7$ MeV used previously.

Although the schematic representations in Figure 2-6 are delineated clearly, one would expect quantum-mechanical exchange to blur the individual components of the nucleons. This would show up in the magnetic moments, since that parameter is known to be very sensitive to structural detail. If, for example, the muon retained its free-space characteristics, the magnetic moment of the proton would be the same -1.0 magneton of the muon, or -8.88 nuclear magnetons (μ_N) when normalized to the mass of the proton by the factor $m_p/m_\mu = 8.88$. This can be predicted because none of the spinless pions would contribute to the magnetic moment.

However, if only one third of the muon remains to contribute to magnetic effects, and the remaining two thirds of its angular momentum are equally divided among the two pions, then the net magnetic moment would be $+2.82$ μ_N, as calculated in Appendix C. This is very close to the observed value of $+2.79$ μ_N.[3]

The comparison is more complicated in the case of the neutron. Two possible analyses of the neutron are considered in Appendix C. The first assumes that the $-\frac{1}{3}e$ charge and the spin from the lost muon segment are evenly shared between the two surviving segments of the π^0. In the second, the charge and spin are constrained to one segment of the π^0, presumably the one furthest from the muon segment because of electrostatic repulsion. In the first case, the magnetic moment of the neutron works out to be -1.19 μ_N; in the second, -2.09 μ_N. These values compare favorably with the measured value of -1.91 μ_N. Due to the closer match in the second case, it is likely that this structure more accurately represents the neutron. However, the actual situation may be a weighted combination of the two states.

It should be mentioned that the original version of the standard model in 1961 assigned spin values of $\frac{1}{2}\hbar$ to all quarks. Since the magnetic moments of the nucleons are used to calculate the moments of the quarks, the quarks cannot themselves be used to assess this assumption. (See Appendix C for more on this subject.) Initial deep inelastic scattering with protons seemed to bear out the value of $\frac{1}{2}\hbar$, and this was accepted for 25 years. However, as measurements were made at higher energies, it became apparent that the presumed quarks could only account for 25-30% of the proton's spin. This became known as the *spin crisis*: quarks do not seem to be the spin-½ particles first assumed by Gell-Mann and Ne'eman.[5,6]

There is much to be said about the straightforward way in which the resonance model deals with the previously intractable problem of nucleon magnetic moments. It is also important to note that it can explain the spin crisis. If the proton is viewed as an assemblage of fractionally charged segments, then the five segments that make up the central cluster would spin, while the eighteen segments of the hexagonal pion ring would not. The ratio of spinning to non-spinning units would then be 5 : 18, or 28%, well within the range of values observed in the scattering experiments.

2-IV. THE TRANSITORY SPIN-½ BARYONS

The other, heavier spin-½ baryons decompose into nucleons after very brief lifetimes, on the order of 10^{-10} second (10^{-20} second in the case of the Σ^0).[1] These transitory particles seem to correspond to nucleons with extra pions in their outer rings. For example, the Σ^+ may be equivalent to a proton with an additional $\pi^+\pi^-$ pair, changing the hexagonal pion ring into an octagon. Similarly, the Σ^0 may be equivalent to a neutron with an additional $\pi^+\pi^-$ pair. It is also reasonable to suppose that the Σ^- is formed by the addition of a π^- and a π^0 to a neutron. The schematic diagrams in Figure 2-7 illustrate the proposed structures and the relationship of the sigma baryons to the nucleons. The masses, when calculated by the same means as the nucleons, fall within a fraction of a percent of their observed values. (See Table V.)

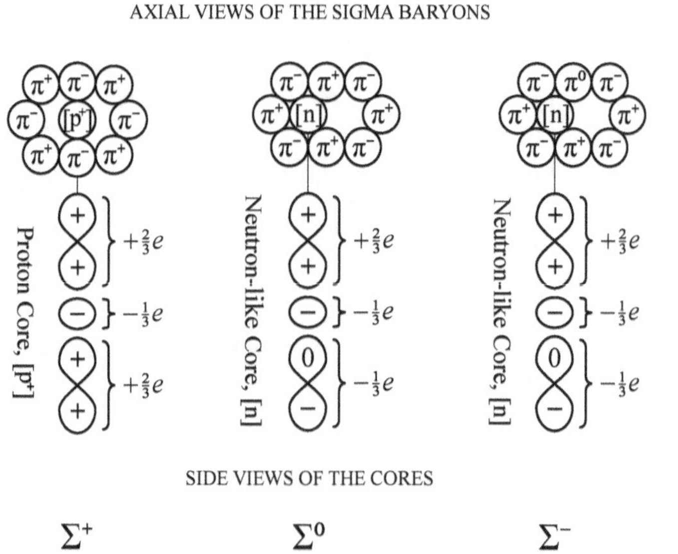

AXIAL VIEWS OF THE SIGMA BARYONS

SIDE VIEWS OF THE CORES

$$\Sigma^+ \qquad\qquad \Sigma^0 \qquad\qquad \Sigma^-$$

Figure 2-7. *STRUCTURES PROPOSED FOR THE SPIN-½ SIGMA BARYONS.*

In the axial views, top, proton cores and 'neutron-like' cores are labeled [p⁺] and [n]. The neutron-like cores seem to differ from true neutrons in having spin distributions that are more evenly shared between the two bottommost segments. The $-\frac{1}{3}e$ charge may be similarly distributed.

The positively charged core of the Σ^+ is expected to draw the negative pions of its ring inward. Whereas there appears to be room for an additional pion inside the rings of the Σ^- and the Σ^0, the more compact Σ^+ may preclude this.

It should be cautioned that these structural choices are not the only possibilities. The Σ^-, for example, could perhaps be equivalent to a proton with two additional π^-s in the outer ring. This alteration would have little effect on the calculated mass, but it is disfavored for two reasons:

1) the addition of two π^-s to the proton would be unfavorable from an electrostatic point of view, and

2) such a composition would be inconsistent with the predominant (nearly 100%) decay of the Σ^- into a neutron plus π^-. In contrast, the Σ^- proposed in Figure 2-7 reveals these decomposition products within its structure.

Similarly, the Σ^+ might be based on a neutron-like core, like the Σ^-, but with opposite charges in the pion ring. Its two nearly equal decay pathways might suggest that the two structures exist as variant forms of the Σ^+. However, the positive magnetic moment of the Σ^+ is incompatible with a neutron-like core.

Since the eight-pion rings of the sigma baryons are larger than the six-pion rings of the nucleons, their bonding to the pions in the central cluster should be weaker than the 15 MeV of the nucleons. For the Σ^+, this would mean an increase in the effective mass of the two $+\frac{2}{3}e$ segments in the core. *Zero* bonding would lower the magnetic moment to $+1.94$ μ_N. (See Appendix C.) Presumably, an intermediate binding energy in the Σ^+ would result in a magnetic moment somewhere between this and the $+2.79$ μ_N of the proton. Indeed, the Σ^+ moment is found to be $+2.46$ μ_N.[9]

The Σ^- can also be rationalized at a qualitative level to support Figure 2-7. With weaker bonding between the central pions and the pion ring than in the neutron, it is possible that the $\frac{1}{3}\pi^{(-)} + \frac{1}{3}\pi^{(0)}$ subassembly in the core could spin as a unit, rather than the spin being restricted to the outer $\frac{1}{3}\pi^{(-)}$ segment as in the neutron. (See Section 2-III.) This arrangement is examined in Appendix C, and found to predict a moment of -1.19 μ_N. This is very close to the observed -1.16 μ_N.[9]

While the 7.4×10^{-20}-second lifetime of the Σ^0 is too brief for its magnetic moment to have been measured, the similarity of its proposed core to that of the Σ^- suggests that their magnetic moments should be comparable.

The Ξ^0 structure shown in Figure 2-8 follows this sequence by adding a π^0 to the central gap of the Σ^0. This assignment was guided by the decay of the Ξ^0, which principally produces a Λ plus a π^0. Furthermore, the Ξ^0 magnetic moment of -1.25 μ_N is quite close to that of the Σ^-, which has a proposed core composition similar to the Σ^0.

The structure of the Ξ^- is suggested by the fact that its magnetic moment is negative, like the n, Σ^-, and Ξ^0; but is substantially weaker. This implies that the added π^0 may spin with the $\frac{1}{3}\pi^{(0)} + \frac{1}{3}\pi^{(-)}$ subassembly of the Ξ^- core to form a $\frac{4}{3}\pi^0 + \frac{1}{3}\pi^-$ spin unit. In Appendix C, this scenario is found to produce a magnetic moment of -0.53 μ_N, encouragingly close to the measured value of -0.65 μ_N.[9] Note that there would be no corresponding Ξ^+ if, as it seems, there is no room in the Σ^+ for additional pions.

AXIAL VIEWS OF THE XI BARYONS

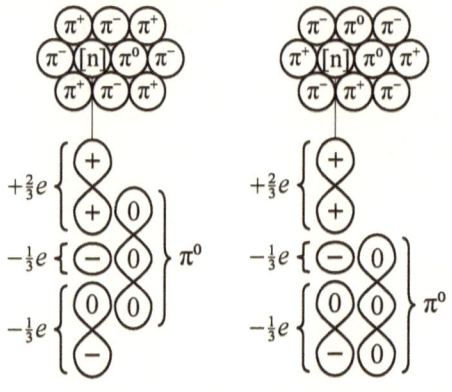

SIDE VIEWS OF THE CORES

$$\Xi^0 \qquad\qquad\qquad \Xi^-$$

Figure 2-8. *STRUCTURES PROPOSED FOR THE SPIN-½ XI BARYONS.*

These are derived from the Σ^0 and Σ^-. There are only two xis, perhaps because only the two sigmas can accommodate an additional pion.

The intriguing lambda completes the baryon octet. It is believed to have the same fundamental components as the Σ^0, but arranged in a substantially different structure, as shown in Figure 2-9. This structure is formed when a third annihilation occurs between a segment of a π^+ in the ring and the neutron-like core. The outer $-\frac{1}{3}e$ segment of the core is lost to this reaction, and a π^- bonds to the surviving portion of the π^+, reducing the ring to the hexagonal form of the nucleons. This accounts quite well for the mass difference between the Λ and the Σ^0. Also, the magnetic moment of the Λ, -0.613 μ_N,[9] is close the Ξ^-. This suggests that all five segments in the lower subassemblies of the Ξ^- and Λ share the spin of the lost muon segment.

Of the three sigmas, why should only the neutral meson collapse into such a structure? The Σ^+ would end up with three positive elements in its core, an electrostatically unfavorable situation. Perhaps Σ^- collapse might not be favorable either, if the smaller hexagonal ring is only able to bear positive and negative pions in equal numbers.

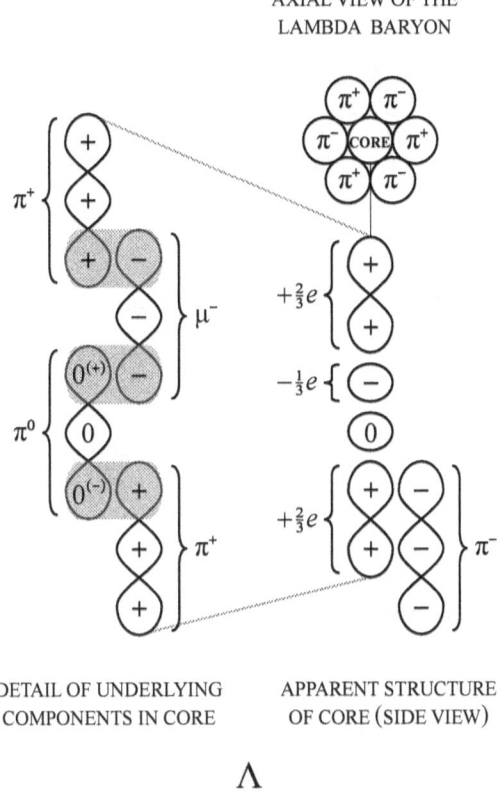

AXIAL VIEW OF THE
LAMBDA BARYON

DETAIL OF UNDERLYING
COMPONENTS IN CORE

APPARENT STRUCTURE
OF CORE (SIDE VIEW)

Λ

Figure 2-9. *STRUCTURE PROPOSED FOR THE SPIN-½ LAMBDA BARYON.*

The underlying components of the core at left do not include the additional π^- shown on the right. As this structure suggests, this composite neutral particle has an antiparticle, $\overline{\Lambda}$, like the composite neutron but unlike the elementary neutral pion.

The masses of the transitory spin-½ baryons derived from the structures shown in Figures 2-7 through 2-9 are summarized in Table V. All fall within 1% of their observed values. As expected, the calculated masses of the Σ^- and Σ^+ are less than observed, since no correction was made for the likely reduction in binding energy between their core and ring pions. Adjustment factors reflecting such effects would be in the spirit of the semi-empirical mass formulae used for atomic nuclei.[1]

The accuracy of the calculated masses provides significant support for assigning these composite structures to the baryons. No other combination of resonances was found to match the observed masses so closely, or to correlate with so complete a set of baryons.

TABLE V. *Transitory Baryons of the Spin-½ Octet.*

The bullet ‹•› indicates an additional segmental annihilation in the core. *

Particle	Proposed Composition	Calculated Mass (MeV)	Observed Mass (MeV)	Discrepancy
Σ^+	$p^+ + \pi^+ + \pi^-$	1187.4	1189.4	−0.16%
Σ^0	$n + \pi^+ + \pi^-$	1188.7	1192.6	−0.3%
Σ^-	$n + \pi^- + \pi^0$	1187.1	1197.4	−0.9%
Ξ^0	$\Sigma^0 + \pi^0$	1315.6	1314.8	+0.06%
Ξ^-	$\Sigma^- + \pi^0$	1320.4	1321.3	−0.7%
Λ	$•\Sigma^0$	1110.6	1115.7	−0.5%

While the standard and resonance models usually provide reasonably consistent results, there is a major discrepancy between their predictions for the magnetic moment of the Σ^0. The standard model predicts +0.79 μ_N. (See Appendix C.) In contrast, the resonance model predicts a magnetic moment close to the −1.16 μ_N of the Σ^-, as illustrated in Figure 2-7.

Although the structures shown in Figure 2-7 are not the only possibilities for the sigmas, there does not seem to be any structure that could bear a magnetic moment anywhere near the standard model's +0.79 μ_N. This is because the positively charged components of the central cluster maintain charges of $+\frac{2}{3}e$. If two such components are present, as in the p^+ and the Σ^+, the magnetic moment should be at least +1.91 μ_N. On the other hand, if only one such component is present, as is the case for the n, Σ^-, Ξ^0, Ξ^-, and Λ, the magnetic moment should be negative. Thus even a measurement of the *sign* of the Σ^0 moment may distinguish between the models.

---

* Figure 2-9 shows a π^+ knocking out the $\pi^{(-)}$ segment for the third annihilation in the Λ. However, since the core pion absorbs the spin of the muon segment it destroys, it takes on the electromagnetic binding effects of that segment, and should behave somewhat like a muon. Thus for this third annihilation we have used an approximation of $\pi•\mu$ interference rather than $\pi•\pi$ interference.

2-V. THE SPIN-³⁄₂ BARYON DECUPLET

For a complete representation of the baryons, the resonance model needs to be applied to the spin-³⁄₂ baryon decuplet in Figure 2-5, that was first explained in 1961 by invoking $SU(3)$ symmetry.[1] With such an extension, the resonance model will exceed the standard model in scope, as it provides the masses and magnetic moments of the baryons.*

Although the spin-³⁄₂ baryons are expected to be substantially more complex than their lighter spin-½ cousins, the resonance model suggests tentative descriptions of them. The details are less certain than for the spin-½ resonances, due both to the complexity of the spin-³⁄₂ baryons and the paucity of magnetic-moment data available for them. Of the entire decuplet, only the rather exceptional Ω^- lives long enough for its moment, $-2.02 \, \mu_N$, to have been well determined.[9]

Nevertheless, a good fit to the delta baryons was found by assuming that each particle has an outer hexagonal ring of pions, as do the nucleons, but with a compound central cluster consisting of two nucleon cores plus a muon, as illustrated in Figure 2-10. Each of the three core components contributes a spin of ½ to the baryon.

There are four possible electronic-charge values of various combinations of the central components. They can be identified with the four delta baryons listed in Table VI.

TABLE VI. The Delta Baryons.

[p] and [n] represent the central clusters of the proton and the neutron.

Particle	Composition of Central Cluster	Calculated Mass (MeV)	Observed Mass (MeV)	Discrepancy
Δ^{++}	$[p] + [n] + \mu^+$	1234.1	1230.8	+0.3%
Δ^+	$[n] + [n] + \mu^+$	1232.5	1231.5	+0.08%
Δ^0	$[p] + [n] + \mu^-$	1234.1	1233.5	+0.05%
Δ^-	$[n] + [n] + \mu^-$	1232.5	1232 ± 2	+0.04%

* With the addition of the charm quark, c, the two dimensions of Figure 2-5(b) grow to three, for a vigintuplet (20-plet) of triplet permutations of the u, d, s, and c quarks. Many of the expected particles, such as (ucc)' associated with a Ξ_{cc}^{++} and (ccc) associated with an Ω_{ccc}^{++}, have not been observed. The addition of the bottom and top quarks, b and t, further complicates the picture. For a discussion of recent developments, see *QUARK MODEL*, revised by C. Amsler and C. G. Wohl.[9]

This set is presumably limited to four because the combination [p] + [p] + μ⁺ is electrostatically unfavorable. For simplicity, only the more compact of two possible Δ⁺ structures is shown. The other is [p] + [p] + μ⁻. There is no combination that will produce a charge of −2e.

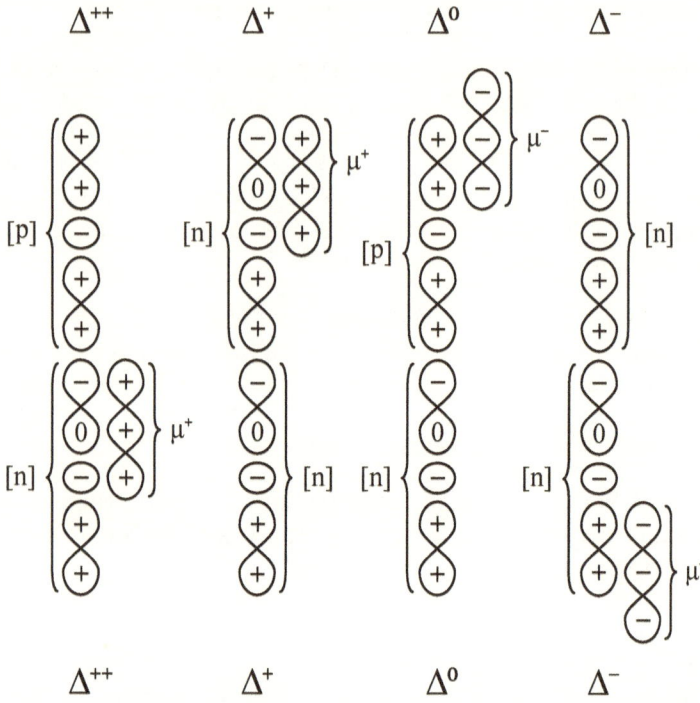

Figure 2-10. CENTRAL COMPOUND CLUSTERS of the four spin-³⁄₂ delta baryons, as proposed in the text. The clusters consist of three components, each with a spin of ½. The muons are situated to maximize electrostatic binding energy while minimizing distortion to the outer hexagonal ring of pions (not shown).

The spin-³⁄₂ sigma-star baryons might be equivalent to deltas with two extra pions, for octagonal outer rings similar to those of the spin-½ sigmas. However, in order to match the observed masses, it is necessary to assume that one of these pions annihilates a segment of the unassociated muon in the core. This brings the number of segmental annihilations in the sigma-stars to five. The extra 82 MeV of binding energy would also help stabilize the particle, increasing its lifetime. (See Appendix E.) Tentative structures for these baryons are summarized in Table VII.

Presumably a Σ^{*--}, equivalent to $\Delta^- + \pi^- + \pi^0$, would be electrostatically unfavorable: The $[n]+[n]+\mu^-$ core of the Δ^- has negatively charged sections at both extremities, which would hinder the addition of another negative particle such as a π^-. Nor could the pion be added to the ring, because then two π^-s would be forced to be adjacent. (See Figure 2-10.)

Three particles remain to complete the spin-$\frac{3}{2}$ decuplet: the Ξ^{*0}, Ξ^{*-}, and Ω^-. Although the details of these complex particles remain to be developed, the calculated mass values in Table VII fit rather well.

TABLE VII. *The Remaining Baryons of the Spin-$\frac{3}{2}$ Decuplet.*
A bullet ‹•› indicates that the pion annihilates a segment of the free muon.

Particle	Proposed Composition	Calculated Mass (MeV)	Observed Mass (MeV)	Discrepancy
Σ^{*+}	$\Delta^+ • \pi^- + \pi^+$	1398.6	1382.8	+1.1%
Σ^{*0}	$\Delta^0 • \pi^+ + \pi^-$	1400.5	1383.7	+1.2%
Σ^{*-}	$\Delta^+ • \pi^- + \pi^-$	1398.6	1387.2	+0.8%
Ξ^{*0}	$\Sigma^{*+} • \pi^- + \pi^0$	1548.3	1531.8	+1.1%
Ξ^{*-}	$\Sigma^{*0} • \pi^- + \pi^0$	1549.2	1535.0	+0.9%
Ω^-	$\Xi^{*0} + \pi^-$ (?)	1656.3	1672.45	−1.0%

The xi-stars are equivalent to sigma-stars with an additional π^- and π^0. The mass calculations assume segmental annihilation between the π^- and the third muon, bringing the number of annihilations to six. Thus the calculated mass of the Ξ^{*-} is the mass-sum of the Σ^{*0}, π^-, and π^0; less 15 MeV binding energy for the π^-, 12 MeV for the π^0, and 82 MeV in additional interference. This extra binding energy stabilizes these particles in comparison to the sigma-stars. (See Appendix E.) Since the third muon has now fused with two pions, the centers of the xi-stars are analogous to triplet nucleon cores. There are eight pions left in the outer ring, similar to the spin-$\frac{1}{2}$ xi baryons.

The meson and baryon masses calculated to this point all fall within 1% or so of their measured values. The positive discrepancies in the sigma-stars and xi-stars may be due to additional bonding possibilities in their rather complex structures. See Appendix E for how these proposals are generally borne out by the baryons' decay products. If the structures seem complex, remember that these are highly unstable particles which decompose within 10^{-23} second or so.

The mass of the omega baryon is similar to a xi-star with an additional pion, and its negative magnetic moment implicates the Ξ^{*0}, with its [n][n][p̄] core, in particular. The structure proposed in Figure 2-11 has a core that is considerably more compact than elongated central clusters of the deltas, and therefore possibly more strongly binding and more stable. Nonetheless, the omega lifetime is still remarkably long, on par with the spin-½ baryons.

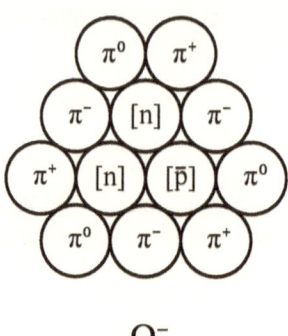

Ω^-

Figure 2-11. *AXIAL VIEW of one structure tentatively suggested for the omega baryon. The three central clusters are the core components of a Δ^+, with the third muon modified by two additional π^-s to form what is effectively the core of an antiproton, $\pi^- \mu^+ \pi^-$, indicated [p̄] in the figure.*

2-VI. CONSERVATION OF STRANGENESS

The word *strangeness* was coined to describe a somewhat mystical property. David Griffiths describes the development of the term:

> It is some measure of the surprise with which these new heavy baryons and mesons were greeted that they came to be known collectively as "strange" particles. ... Not only were the new particles unexpected; there is a more technical sense in which they seemed "strange": They are *produced* copiously (on a time scale of about 10^{-23} sec), but they *decay* relatively slowly (typically about 10^{-10} sec). This suggested ... that the mechanism involved in their production is entirely different from that which governs their disintegration. In modern language, the strange particles are *produced* by the *strong* force, ... but they *decay* by the *weak* force. ... [Theorists later] assigned to each particle a new property (Gell-Mann called it "strangeness") that ... is conserved in any strong interaction, but ... is *not* conserved in a weak interaction.[10]

The strange quark was later postulated to account for this property. However, strangeness can be rationalized without the need for a quark.

Once a charged particle forms from a standing-wave resonance, it may move sufficiently fast relative to another particle that their collision could generate additional particles. An observer in the center-of-mass reference frame would see the characteristic dimensions of the moving particles, their resonant wavelengths, reduced by Lorentz contraction. If the relative wavelengths were sufficiently short, other particles consistent with conservation of energy and momentum could emerge from the field spawned by the collision.

Addressing proton-antiproton collisions in this manner provides a phenomenological explanation for the observed conservation of strangeness.[11] One need only assume that when the kinetic energy of the collision is sufficiently great, it is possible to replicate entire pion rings *en masse*. (Producing the more massive protons would require higher energies.)

Two of the pions in such a nascent hexagonal ring could combine with a neutron to form a Σ^0 (strangeness -1), while the remaining four would coalesce to form a K^0 (strangeness $+1$), thereby conserving strangeness. The formation of a Ξ^0 (strangeness -2) would require the addition of a π^0 to the Σ^0. However, if collisions frequently replicate the pion rings of the colliding particles, there would be five pions left after the formation of the Ξ^0. Four would form a K^0 (strangeness $+1$), while the fifth would remain unassociated (strangeness 0). Again, strangeness is conserved.

From this perspective, conservation of strangeness may be a product of geometry rather than being required by some abstract quantum relationship.

2-VII. HEAVIER PARTICLES

In recent years, a limited number of more massive particles has been observed. These include the J/ψ, bottom (B), and upsilon (Υ) mesons, summarized in Table VIII.[4]

These particles were evaluated as possible third-generation composites with nearly immediate positive results. For example, the J/ψ may be a $\tau^+\tau^-$ pair circulating in a spin-1 quasi-ground state. In this case the binding energy would be $m_{J/\psi} - 2m_\tau = 457$ MeV. Such exceptionally large binding energies are expected for third-generation particles, as explained in Chapter 1.

Further support for this assignment comes from the fact that the J/ψ exhibits energy levels with S, P, and D states, suggestive of the hydrogen atom and positronium, but with much larger steps in energy—on the order of 600 MeV.[12] Indeed, the observed S state includes a true spin-0 ground state at 117 MeV below the J/ψ that has been designated the eta charmed meson, η_c.[12] Again, this is suggestive of positronium. (See Figure 1-5.)

TABLE VIII. *The Third-Generation Composite Mesons.*

The spins of the bottom mesons have not been measured, but are predicted to be zero by the standard model.

Particle	Rest Mass (MeV)	Electronic Charge	Spin	Lifetime (sec)*
J/ψ	3096.9	0	1	0.8×10^{-20}
$B^\pm$	5279.0	$\pm e$	0?	1.7×10^{-12}
$B^0, \bar{B}^0$	5279.4	0	0?	1.5×10^{-12}
Υ	9460.3	0	1	1.3×10^{-20}

The bottom mesons, B, may be the third-generation analogues of the second-generation kaons: that is, four D_s mesons in a close-packed cluster (assuming the D_s is correctly identified as the 3rd-generation meson). The B^0, for example, may be a composite of two D_s^+ and two D_s^-. Its binding energy would be about 2595 MeV, allocated equally to the four $D_s^+ D_s^-$ contact points at 649 MeV per bond. This is substantially less than the magnitude of the 840-MeV bond of the $D_s^+ D_s^-$ representation of the J/ψ.[†] Qualitatively, one would expect the B^0 bonds to be somewhat weaker, due to electrostatic repulsion along the diagonals of the cluster. (See the analogous K_{LONG}^0 in Figure 2-2.) Interestingly, recent work, published since this association was made in the resonance model, suggests that the neutral bottom mesons exhibit a **CP** violation similar to that of the neutral kaons.[14]

Analogously, the spin-1 upsilon meson may consist of two circulating B^0s. This would involve two more D_s^+ bonds than the separate B^0 structures. The rest mass of such a Υ is expected to be $2m_{B^0} - 2BE_{D_s^+ D_s^-} = 9262$ MeV. This is reasonably close to the observed rest mass of 9460.3 MeV, supporting the

---

* The lifetimes, τ, of the J/ψ and Υ mesons have been derived from their linewidths, Γ, by the relation $\tau = \hbar/\Gamma$. [13]

† The J/ψ, like other high-order resonances, has a mixed nature. Its decay modes include both leptons ($e^+ e^-$, $\mu^+ \mu^-$) and hadrons ($2\pi^+ 2\pi^- \pi^0$, $3\pi^+ 3\pi^- \pi^0$, ...). Bearing in mind the relation $D_s = \frac{1}{2}\tau^\uparrow + \frac{1}{2}\tau^\downarrow$ from Chapter 1 (*cf.* also Equation 4-4), the spin components of the J/ψ may flip between a weaker 457-MeV $\tau^+ \tau^-$ lepton bond and a stronger 840-MeV $D_s^+ D_s^-$ hadron bond. The characteristics of the J/ψ are likely to be more heavily influenced by the stronger hadron binding energy, since this is the lowest energy state, and indeed hadron decay occurs 88% of the time.

identification of the Υ as a composite third-generation particle, rather than the fourth-generation di-fermion tentatively suggested in Chapter 1.*

2-VIII. CONCLUSION

The elementary particles identified in Chapter 1 can be combined into efficiently packed structures whose rest masses correspond to the kaons and known baryons, as well as to heavier mesons such as the B, J/ψ, and Υ.

Simple binding-energy rules for the elementary particles can establish the masses of the kaons and baryons to within about 1% of their observed values. This falls short of a confirmation of the proposed structures. Nevertheless, the resonance model appears unique in its accuracy and completeness. The validity of these proposals may be tested by the negative magnetic moment they predict for the sigma-zero baryon, which has yet to be measured and which is predicted to be positive by the standard model.

It has often been noted that with a sufficient number of parameters one can fit any set of experimental data. However, this seems a more fitting criticism of the standard model, with its 21 or so arbitrary parameters, than of the resonance model, which requires only c, e, $\hbar$, and ω_0, all of which are of fundamental significance.

---

* The base masses of the third and fourth generations differ by a factor of four, making it difficult to distinguish elementary fourth-generation particles from simple third-generation composites.

† *Endnote from Figure 2-3(a)*

The two neutral K mesons shown in Figure 2-3(a), K^0 and $\bar{K}^0$, are related to the K^0_{LONG} and K^0_{SHORT} by a linear transformation,

$$K^0_{SHORT} = \sqrt{2}\left[(1 - \varepsilon)K^0 + (1 + \varepsilon)\bar{K}^0\right]$$

$$K^0_{LONG} = \sqrt{2}\left[(1 + \varepsilon)K^0 + (1 - \varepsilon)\bar{K}^0\right],$$

where $\varepsilon \approx 2 \times 10^{-3}$ is the amount of mixing.[3]

The reason for this complexity, according to the standard model, is that the neutral kaons take on different characteristics depending on whether the strong or weak force is governing a particular interaction. The K^0 and $\bar{K}^0$ are viewed by the standard model as eigenstates of the strong interaction which produces them and which conserves strangeness. However, the decay of the neutral kaons is reasonably well governed by conservation of Charge and Parity (CP) reversal. Only a small deviation from CP invariance has been observed, represented by ε in the equations above. Yet this small factor could have major consequences. If CP invariance is at all directly violated, as has been reported but not confirmed, there should be a measurable difference in the branching ratios of the neutral kaons. This difference could, in principle, be communicated to physicists in a distant galaxy, who could then perform an experiment to determine whether our galaxy were made out of matter or antimatter![3]

The resonance model allows no such distinction. If the pions within the neutral kaons were replaced by their antiparticles—that is, $\pi^+ \leftrightarrow \pi^-$, with π^0 unchanging—the result would be indistinguishable from the original. This specifically prohibits using the branching ratios of the neutral kaons to distinguish matter from antimatter, a hypothesis that can be tested with present-day equipment.

ADDENDUM TO CHAPTER 2

Figure 2-6(b) suggests that the neutron might have an electric dipole moment due to the charge separation between the $+\frac{2}{3}e$ charge segment at the top of the core and the two $-\frac{1}{3}e$ segments at the bottom. Now it is well known that the existence of an electric dipole moment in an elementary particle would violate both time reversal, T, and a parity operation, P.[15] Since the Standard Model predicts that the neutron should not have an electric dipole moment either, considerable effort has been put forth to verify this. To date it has been determined that any moment must be less than 0.63×10^{-25} $e \cdot$cm.[9] This is much smaller than a charge of $\frac{2}{3}e$ multiplied by the 10^{-13}-cm diameter of the neutron that would be expected from the schematic in Figure 2-6(b). The resolution of this apparent discrepancy is that the figure represents an instant frozen in time. It possible that the charge segments in the core of the neutron are not stationary but circulate rapidly along the axis—at nine times the speed of light per Equation (1-5b). If this is the case, then when measured at low frequencies the electric dipole moment would average out to zero. This would be true of the other baryons discussed in Chapter 2 as well.

[1] G.D. Coughlan and J.E. Dodd, *The Ideas of Particle Physics*, 2nd ed. (Cambridge University Press, 1991), 59-62.

[2] R.B. Leighton, *Principles of Modern Physics* (McGraw-Hill, New York, 1959), 549-551.

[3] Coughlan and Dodd, 77-79.

[4] *Ibid.*, 233-240.

[5] K. Rith and A. Schafer, 'The mystery of the nucleon spin.' *Scientific American* (July 1999), 58-63.

[6] B. Schwarzschild, 'Experiments at Jefferson Lab and MIT probe for strange quarks in the proton.' *Physics Today* (June 1999), 21-23.

[7] George Johnson, *Strange Beauty: Murray Gell-Mann and the Revolution in Twentieth-Century Physics* (Alfred A. Knopf, New York, 1999).

[8] L.M. Lederman and D.N. Schramm, *From Quarks to the Cosmos* (Scientific American Library, New York, 1989), 100-111.

[9] K. Hagiwara *et al.*, 'The Review of Particle Physics.' *Physical Review*, **D66** (2002). Available online at *http://pdg.lbl.gov*.

[10] D. Griffiths, *Introduction to Elementary Particles* (Harper and Row, New York, 1987), 32.

[11] Leighton, 651-661.

[12] Coughlan and Dodd, 154-158.

[13] Gordon Kane. *Modern Elementary Particle Physics: The Fundamental Particles and Forces* (Addison-Wesley, New York, 1993), 119-122.

[14] James Glanz, 'A second hint of symmetry violation.' *Science* (18 December 1998), 2169.

[15] Donald H. Perkins, *Introduction to High Energy Physics*, 4th ed. (Cambridge University Press, 2000), 81-86.

Chapter 3

MASS, GRAVITATION, AND COSMOLOGICAL IMPLICATIONS

The concept of mass arises naturally from the resonance model, which is thus well placed to bring gravitation into a unified theory. It further suggests that the big bang began as a single primordial entity, analogous to the standing-wave resonant particles described in Chapter 1, but on the order of 4×10^{21} segments. The rapid decomposition of such an entity provides a specific scenario for inflation, galaxy formation, and, perhaps, the production of cold dark matter.

3-1. INTRODUCTION

The first two chapters of this monograph present a case that all massive particles can be described as standing-wave resonances of the electromagnetic field. This chapter will show that the concept of inertial mass follows from such resonances as well.

Although gratifying, this result leaves unanswered the basic question of the origin of mass and charge. A unified theory should offer a convincing portrayal of creation. While this remains a tall order, a number of the concepts covered in this chapter may prove useful.

One additional postulate is needed, that before the beginning of our universe there was nothing: no energy, no gravitation, no, no space, no electronic charge. Bridging the presumed nothingness prior to the big bang with what followed requires the sort of *zero-sum* universe favored by Guth,[1] Hawking,[2] Tyron,[3] and others,[4] in which all energy contributions add to zero and which 'costs nothing' to create. Heinz Pagels, who struggled with this issue, expressed his thoughts most eloquently:

The nothingness 'before' the creation of the universe is the most complete void that we can imagine—no space, time or matter existed. It is a world without place, without duration or eternity, without number—it is what mathematicians call 'the empty set'. Yet this unthinkable void converts itself into the plenum of existence—a necessary consequence of physical laws. Where are these laws written into that void? What 'tells' the void that it is pregnant with a possible universe? It would seem that even the void is subject to law, a logic that exists prior to space and time.[5]

The world view of the resonance model suggest an innovative zero-sum scenario for creation.* This includes a plausible hypothesis for how mass and charge could arise from essentially nothing, and covers the inflationary period, the era of galaxy formation, and the final phase of creation that lead to the preponderance of matter over antimatter.

The most elusive component of any unified theory is gravitation. This will be addressed both before and after the discussion of creation, as certain gravitational concepts are required to appreciate the creation model, and *vice versa*. Ending the chapter is a discussion of cold dark matter, which many cosmologists believe accounts for 90% or more of the mass of the universe. Suggested as a candidate for this matter is a hypothetical particle, similar to the neutron but stable and substantially more massive, that is presented along with possible methods for its detection.

3-II. THE ORIGIN OF INERTIAL MASS

The basic premise of the resonance model is that all massive particles are products of certain allowed standing-wave resonances of the electromagnetic field. These particles have a polar axis in their inertial reference frame along which electromagnetic energy circulates. The field travels an odd number of half wavelengths along the polar axis and then returns in the opposite direction.

So long as such a particle remains at rest, no work is needed to maintain its position. However, if a force **F** is applied to one end along the polar axis, the particle will begin to move relative to its inertial reference frame. This means that the circulating electromagnetic energy which started out at the point of application of the force will return after a complete circulation to some position displaced by a distance Δz.

---

* There are also several non-zero sum perspectives. An intriguing model by Sternglass is one.[4]

If the now-reduced circulation path length of the electromagnetic field is to remain resonant, the resonance frequency, ω, must increase to compensate. Thus the particle's mass energy will also increase, as

$$mc^2 = \omega\hbar \tag{3-1},$$

and any deviation from an inertial reference frame will require that work be done on the particle to supply this increase in energy. From the perspective of the original coordinate system, the length of the particle will appear to shrink by $\Delta z/2$, which corresponds to the Lorentz contraction.

Once the applied force is removed, the particle will have been transferred to a new inertial reference frame. Restoring it to its original frame will require an additional expenditure of work. Seen from the original reference frame, the particle is first seen to speed up, increasing in mass, and then to slow down, returning to its original mass.

Since the differential work, dW, done on the particle by $\mathbf{F}$ while moving it a distance $d\mathbf{z}$ is a change of energy, it equates to the change in frequency, $d\omega$:

$$\mathbf{F} \cdot d\mathbf{z} = dW = \hbar\, d\omega \tag{3-2}.$$

The de Broglie* relationship relates momentum, $\mathbf{p}$, to wavelength, λ, by

$$|\mathbf{p}| = \frac{h}{\lambda} = \frac{\hbar\omega}{c} \tag{3-3}.$$

Substituting into Equation (3-2), and recognizing that $dz = c\, dt$,

$$\mathbf{F} \cdot d\mathbf{z} = \hbar\frac{d\omega}{dz}dz = c\frac{d\mathbf{p} \cdot d\mathbf{z}}{dz} = \frac{d\mathbf{p} \cdot d\mathbf{z}}{dt} \tag{3-4},$$

Or

$$\mathbf{F} = \frac{d\mathbf{p}}{dt} \tag{3-5}.$$

This is the general (classical and relativistic) version of $\mathbf{F} = m\mathbf{a}$. Thus this conception of mass satisfies the equations of motion. Note that although this analysis presumed a force applied in the axial direction, $\mathbf{z}$, it is generalizable to all directions of motion. This is because it is always possible to find a Lorentz transformation to an inertial coordinate system in which the direction of the force is collinear with the axis of the particle.

---

* *pronounced "də BROY"*

3-III. GRAVITATION: SUBATOMIC PERSPECTIVE

The resonance model resolves subatomic particle masses in a simple and natural manner. It treats the particles as finite standing waves rather than as infinitesimal points, eliminating the infinite energies associated with the electromagnetic and gravitational fields of singularities.

This eliminates one of the principle difficulties in integrating gravitation with electromagnetism and the nuclear forces. Although it is by no means clear how to proceed at this point, the following considerations bear keeping in mind:

(1) Since the strong nuclear force is identified with electromagnetism in Chapter 1, and the weak force is explained in electromagnetic terms in Chapter 4, it is reasonable to assume that gravitation may also share a common origin with electromagnetism.

(2) Both gravitational and electromagnetic forces have infinite ranges that decrease with the square of distance, again suggesting a connection.

(3) Maxwell's equations for electromagnetism are invariant under a parity operation, yet weak interactions, which involve neutrinos, are not. This issue will be further examined in Chapter 4. For now suffice it to note that this apparent incongruity can be resolved if one presumes that electronic charge is not precisely conserved by weak interactions, but varies minutely between the particle generations introduced in Chapter 1. Any such variation must be extremely small for all stable particles, since highly precise measurements have not revealed any disparity.

(4) The strength of the electrostatic force relative to gravitation is 4×10^{42}. Specifically, the electrostatic attraction between an electron and positron is

$$F_e = K \frac{e^2}{r^2} \qquad\qquad (3\text{-}6),$$

while the gravitational force between the same particles is

$$F_g = G \frac{m_e^2}{r^2} \qquad\qquad (3\text{-}7).$$

Combining, with $K = (4\pi\varepsilon_0)^{-1}$ in the MKS system,

$$\frac{F_e}{F_g} = \frac{Ke^2}{Gm_e^2} = 4.17 \times 10^{42} \qquad\qquad (3\text{-}8).$$

(5) Since the electronic charge e appears squared within this dimensionless value, the square root, 2.04×10^{21}, may be the number of discrete values that electronic charges can assume. If the steps in this range are of equal magnitude, and if each step is associated with a different particle generation, then the total number of generations in our universe should be 2×10^{21}.

This latter premise, that the number of particle generations is some 2×10^{21}, provides another clue for associating the electromagnetic and gravitational forces.

The simplest resonance is the lowest-order particle, the electron, which consists of a single standing-wave segment and an electronic charge of e. At the opposite extreme, the highest-order (primordial) resonance would have had $N - \frac{1}{2}$ standing wavelengths ($2N - 1$ segments) and a probable electronic charge of zero. The generation below that would have had $N - 1\frac{1}{2}$ wavelengths and an electronic charge of e/N. Each succeeding generation would have had one fewer wavelength and a fractionally larger charge.

A simple model to keep track of the changing wavelength and charge values is to associate charge with a number of left- or right-handed 'twists' within the resonance. Each twist may be thought of as a loop in a tight helical path. Thus the ith standing-wave resonance would have $(n_i - \frac{1}{2})$ wavelengths, such that n_i plus the number of twists, t_i, in the circulation path around the resonance equal N. That is,

$$n_i + t_i = N \tag{3-9}.$$

Electronic charge was originally correlated with left- and right-handed twist partly as a convenience, as this allows for two varieties of charge. The concept was also guided by the work of Theodor Kaluza,[6] who associated electronic charge with a cylindrical fifth dimension, and in so doing derived Maxwell's equations of electromagnetism from general relativity.*

It later became apparent that this correlation may have deep significance. Based on the material presented in Chapter 4, right-handed twist can in fact be identified with positive charge, and left-handed twist with negative

---

* Since the four traditional dimensions (x, y, z, t) of space-time have apparently infinite ranges, it seems odd at first that a fifth dimension could be represented by electronic charge, which is a fixed value, e. Addressing this incongruity, string theorists have associated e with a 'collapsed' or 'compact' dimension. The approach taken by the resonance model is to consider the dimension of charge as ranging over a extremely large, but finite, number of 2×10^{21} discrete steps. Perhaps the other four dimensions are similarly large but finite.

charge. Thus high-order particles with net right-handed twist, $t_{i(R)}$, are described as having a nascent positive charge, much less than e, while high-order resonances with net left-handed twist are described as having a nascent negative charge. At any time during the descent from the primordial entity, the net right- and left-handed twist summed over all particles was zero, presuming there was no net twist in the beginning:

$$\sum_i t_{i(R)} + \sum_j t_{j(L)} = 0 \tag{3-10}.$$

Since twist defines electronic charge, Equation (3-10) states that our universe is electrically neutral.

With this groundwork in place, it is possible to relate the gravitational field of a massive particle to the number of its segments, just as the electronic charge is related to its twist. In the case of the electron (a resonance of order one: one segment and $N = 2 \times 10^{21}$ twists), the ratio of electronic to gravitational forces would be N^2, or approximately 4×10^{42}. At the opposite extreme, in a resonance of order $N - 1$, the ratio would be reversed. The gravitational force would be 4×10^{42} times electromagnetic the force!

Historically, the forces would have been equal at the generation midway between these two extremes. If the electromagnetic contribution of a single twist is equated with the gravitational field of a single standing wavelength, then the two forces are comparable even today.

It should be noted that, unlike twist, which is either left- or right-handed and consequently can realize two types of charge, the number of segments is always a positive number. This simple consequence of geometry may explain why, unlike electromagnetic forces, gravitational forces are always attractive. This is discussed further in Chapter 5.

This argument leads to a new premise, complementary to the resonance model, which may be stated as follows:

> *The electronic charge of a standing-wave resonance is proportional to a number of 'twists' within the resonance, such that the sum of the number of internal twists and external wavelengths in the resonance is a universal constant. Furthermore, the gravitational and electromagnetic forces are of the same magnitude when considered on a per-wavelength and per-twist basis.*

Together with the resonance model, this forms the framework for a unified theory of physics. It is used in the following section to explore creation.

3-IV. THE ORIGIN OF THE UNIVERSE

If critical aspects of the resonance model are experimentally confirmed, they will have profound implications for the origin of the universe. Even before such verification can be attempted, curiosity naturally kindles the imagination. This speculative section should be received in that spirit.

The resonance model suggests that the 'nothingness' prior to the big bang could be arbitrarily described, as if by a Fourier transform, by any number of primordial standing resonances, each with a characteristic number of $N - \frac{1}{2}$ wavelengths ($2N - 1$ segments). These primordial entities are mathematically similar to the resonances introduced in Chapter 1, but differ in that they would not have had any explicit electronic charge. Any value of N would have been possible. As it turns out, the value of N that came to define our universe was around 2×10^{21}.*

For reasons which are unclear, but which may be rationalized as an increase in entropy or as quantum fluctuation,[7] this particular entity divided into two or more daughters. These each lost one wavelength by collapsing it—twisting it internally—so that the total number of wavelengths, standing wavelengths plus twists, remained constant. This would have been the moment of creation.

Each generation in the subsequent descent was considerably more populated that the previous one, and the average twist—the source of electronic charge—increased. Both standing waves (massive particles) and traveling waves (massless photons and neutrinos) were produced. For example, emission of a neutrino would increase the twist of a resonance by one-half turn.

Indeed, *existence* as we know it might be defined by the twist generated during this descent from the primordial entity. As the twist increased and electromagnetism separated from gravitation, we could say that the universe was coming into existence.

After 2×10^{21} generations, the descent reached a lower limit with the single-segment electrons. The surviving particles all had about 2×10^{21} twists. Specifically, electrons had N twists; muons, $N - 1$. The electron can be said to be *fully developed* or fully existent, while early-generation particles with smaller quantities of twist can be described as *nascent*. Thus the muons are almost but not quite fully developed: They are one twist short of an electron,

---

* Unlike Pagels, the authors propose that the 'nothingness' before the big bang was not without number. Universes based on values other than our own should be possible.

and should therefore have an electronic charge less than that of the electron by one part in 2×10^{21}. Similarly, the pion should be 1½ twists and the tau two twists short of the electron.*

Because none of the higher-order resonances, including the primordial entity, were stable, they would have released an intense flux of electromagnetic radiation, including neutrinos, in a long cascade that gave birth to an unimaginably large number of daughter resonances. Decomposition and regeneration would have occurred very quickly, and because the daughter resonances were substantially larger than their mothers, according to the analysis in Appendix A, there would have existed a brief period of profound expansion of the universe.

This period can be identified with the hypothesis of inflation. Inflation is increasingly accepted by cosmologists and physicists because it resolves a number of problematic issues.[11] It postulates that the nascent universe increased in size by a factor of approximately 10^{50}, with a doubling time on the order of 10^{-35} seconds, until it reached roughly the size of a baseball (about 10 cm).[12] This prodigious enlargement thus had an average radial expansion rate on the order of 10 centimeters during the 10^{-32} seconds or so of the inflationary period, or 10^{33} centimeters per second! Peebles estimates the speed was even greater, on the order of 10^{35} cm/sec.[13]

Since this is much faster than the speed of light, 3×10^{10} cm/sec, inflation creates a new conundrum: what determined the rate of inflation? The proposed resolutions include such *ad hoc* mathematical assumptions as a cosmological constant or properties of the yet-unconfirmed Higgs field.[14] In this regard the resonance model may be informative, since the phase

---

* Considering the subtleties involved in accounting for the twist of the annihilated muon and pion segments in the proton core, it does not follow that the proton shares this deviation from the electron. Nor, for that matter, has the charge of the hydrogen atom been measured with sufficient precision to rule out the possibility that it might. If there *is* a difference between the absolute values of the electron and the proton, $|q_e|$ and $|q_p|$, it would cause a small repulsion between all matter in the universe that could be significant over cosmological distances.

H. Bondi and R. A. Lyttleton suggested in 1959 that if $|q_e|$ and $|q_p|$ differ by one part in 10^{18}, electrostatic forces could be the cause of the observed expansion of the universe. Shortly thereafter A. M. Hillas and T. E. Cranshaw[8] reported a maximum disparity of $4 \times 10^{-20} e$. Subsequent work[9] has reduced the upper limit to about $10^{-21} e$. While impressive, these results are limited by the assumption that any disparity is the same for protons and neutrons. However, even a smaller effect might explain recent astronomical observations that the rate of expansion of the universe seems to be increasing with time,[10] without the need to introduce any new physics like the *cosmological constant*, a parameter that has no physical basis at present.

velocity of light is not constant between the resonance generations, but increases quadratically. Perhaps this dictated the rate of expansion. Equation (1-6) establishes that the phase velocity within the primordial resonance would have been $N^2 = 4 \times 10^{42}$ times c, assuming N is 2×10^{21}. As the decay process driving inflation progressed, the phase velocity would have ebbed with each generation. When the last generation was reached it would have stabilized at c, ending inflation—and providing the 'graceful exit' that is missing from other proposals.[15]

Because electrons, not positrons, dominate the matter around us, it seems likely that our galaxy, and perhaps a much larger region of the universe, started out as a single massive resonance of left-handed twist (negative charge) sometime during the inflationary era.* Each subsequent step down the decay chain would then have favored the formation of negative resonances, although the net charge of the region would have been nearly zero. If a positive resonance happened to appear in the lead generation, it would have quickly annihilated a negative resonance and returned its energy to the electromagnetic field. The net result would be that negative particles would have reached the lowest generations first, churning out myriads of negative electrons.

Just before the descent terminated, there would have been a brief interval during which positive pions outnumbered negative pions, which had preferentially decayed into negative muons. These conditions would have been ideal for the formation of the nucleons, whose cores are believed to consist of positive pions plus negative muons, as described in Section 2-III.

Exploding numbers of ever-larger resonances would have materialized with high kinetic energies in a densely packed 'soup'. Their effective temperature was thus very high, and thermal collisions would have started them rotating. (Bose condensates require low temperatures, so linearly polarized bosons would not have behaved two-dimensionally under such conditions.)

---

* It may be that the primal decay generated an overall left-handed twist, as suggested in the footnote to Section 5-I:3(c). If this is the case, the entire universe should be made of normal matter. The universe would then have a small net charge, $q_\text{U} = -e$, but this is no more a problem than the fact that it has a (large) net mass. For q_U to be absolutely zero, the first daughters must have had opposite twists, and whole regions of the universe should be made of antimatter.

Now cosmic rays appear to consist almost exclusively of normal matter, but they are relatively local. Even the most energetic rays can only travel some thirty million light-years before suffering severe damping from the 2.7 K cosmic background radiation of the universe.[16] Thus the possibility of distant antimatter galaxies remains open.

Eventually, as the rate of production dropped, a point would have been reached when the resonances separated faster than the voids between them could be filled by new generations of yet larger particles. The meson soup began to 'boil'. This may have been the era that seeded the galaxies and galaxy clusters we observe today.

It is appropriate to mention at this point how little we really know of galaxy formation. In their recent book, *Gravity's Fatal Attraction*, Mitchell Begelman and Britain's Astronomer Royal Martin Rees offer this perspective:

> Galaxies ... are supported against gravitational collapse through the motions of their stars. Beyond this general statement of gravitational equilibrium, no theory analogous to the well developed theory of stellar evolution exists to explain the gross properties of galaxies. Indeed, we do not know why such things as galaxies should exist at all, let alone constitute the most conspicuous large-scale features of the cosmos. Only through diligent observation have we learned that galaxies possess fairly standardized properties, from which we can infer their structures and something about their fates. ...

> The spiral arms that are such a conspicuous feature of some disk galaxies delineate regions where star formation is proceeding with unusual rapidity. The most spectacular arms seem to correspond to some kind of persistent wave pattern in the disk, but there is still no completely satisfactory explanation of what excites and maintains such a wave. ...

> To achieve a full explanation of galaxies will require setting them in a cosmological context—the 'seeds' that developed into galaxies must have been implanted in the early Universe.[17]

If the seeds of galaxies *were* high-order resonances, this might explain a number of galactic features. It is possible that the two predominant galaxy types, elliptical and spiral, derive their characteristics from the still high-order resonances that formed as the meson sea started to boil. Specifically, the most massive resonances, which were also the smallest in size (see Appendix A), had acquired the least angular momentum during the inflationary period. Their small cross sections were not conducive to thermal collisions. Thus they may have given rise to the elliptical galaxies, which exhibit little rotation. This is consistent with the fact that spherical galaxies, a subclass of the ellipticals, are known to be substantially more massive than spirals, typically by a factor of thirty.[18]

The evolution of massive resonances into galaxies would have mimicked the big bang on a smaller scale, with intricate cascades of decay products. Since the self-gravitation of galaxies is sufficient to stabilize their shape over cosmological time, they may preserve a record of their formation.

The opposing arms seen in spiral galaxies are similar in form to the decay jets observed in high-energy subatomic collisions. Abrupt branching points, evoking particle decay pathways, still seem to be visible in the arms of some galaxies, such as the ones shown in Figure 3-1. Perhaps decay pathways and galaxies form in a similar manner. Certainly both must follow similar energy- and momentum-conserving trajectories.

Spiral galaxies were left rotating at the end of the inflationary period, and their jets twisted into the shapes observed today. The results remain most distinctive in smaller spiral galaxies, presumably because there was little inflation in the resonances that seeded them. Recent Hubble photographs of very young galaxies show that many have reasonably well established spiral arms.[19] Thus it seems that the arms did not gradually coalesce from a nearly uniform gas over the eons, as has long been suggested.

For those familiar with the groundbreaking work on galactic rotation by Vera Rubin and her associates,[20] this interpretation may seem a bit shaky. Rubin *et al.* are most remembered for the finding that rotational velocities of large spiral galaxies are approximately constant outside the central bulge. This has become well known as it is an important line of evidence for massive galactic halos of cold dark matter, discussed later in this chapter. The gradient velocities in such galaxies would produce tight coiling of the spiral arms over the age of the universe. Our own stellar neighborhood, for example, is estimated to have made some fifty revolutions about the Milky Way since the galaxy formed.*

While a high degree of wrapping may occur in large spiral galaxies such as the Milky Way or its neighbor Andromeda, Rubin also determined that in smaller, less luminous Sc-type spiral galaxies, rotational velocities increase nearly linearly with distance from the galactic center. That is, the galactic year is similar for most stars. Such galaxies appear to rotate as nearly rigid bodies, with little winding of their spiral arms.

It is thus conceivable that the spiral structures seen in Figure 3-1 preserve all the winding that has occurred in these Sc galaxies since they formed, and that the configurations of their arms are not dramatically different from

---

* Our solar neighborhood has a galactic year of some 250 million years, about 2 % the estimated age of the universe.[22]

Figure 3-1. BRANCHING STRUCTURE *of a typical SC galaxy,* NGC *3486.*
Lines are centered along the major spiral arms, tracing the presumed
pathways of the resonance cascade that may have formed this galaxy.

BELOW: *With the curvature removed, the branching pattern is reminiscent*
*of particle decay jets. (Photo courtesy of J.-C. Cuillandre/*CFHT.[21]*)*

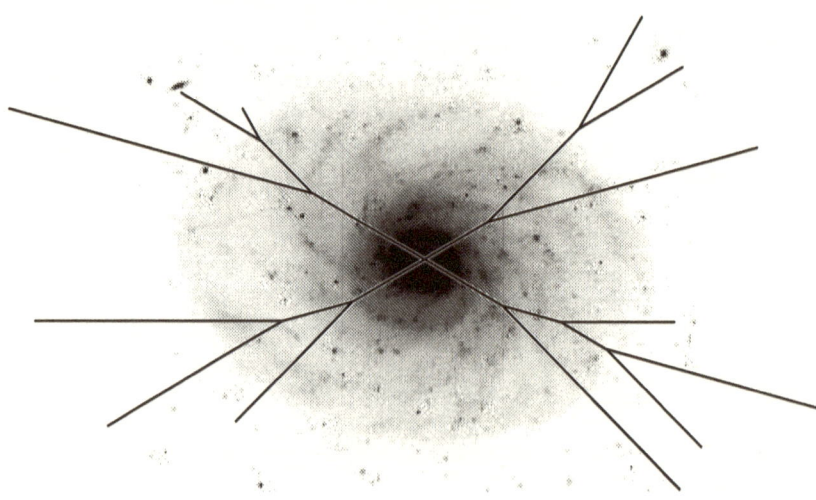

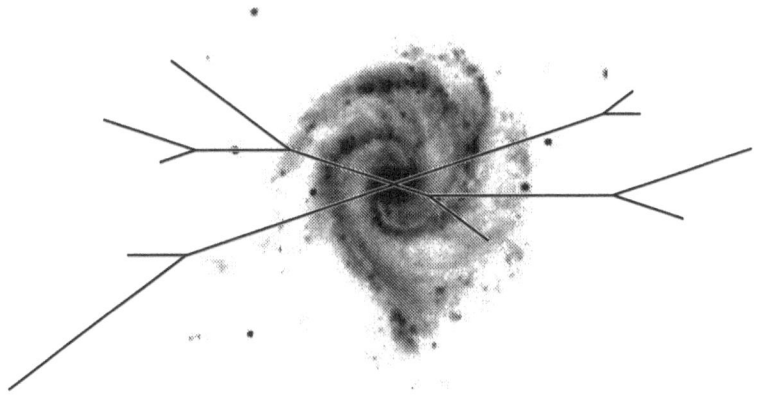

Figure 3-1. *(CONTINUED)* *The structure of galaxy* NGC *5426.*
*(Photo courtesy of J.-C. Cuillandre/*CFHT.*)*

what they were then. Of course, such scenarios allow for later magneto-dynamic evolution such as the formation of galactic density waves.*

3-V. GRAVITATION: COSMOLOGICAL PERSPECTIVE

Section 3-III discussed gravitation in the context of subatomic particles and the electromagnetic forces between them. It did not tackle the magnitude of the gravitational constant, G, in

$$F_g = G \frac{m_i m_j}{r_{ij}^2}$$
(3-11),

where r_{ij} is the distance between masses m_i and m_j. This requires the guidance of cosmology.

The gravitational constant might be construed from the design of the universe, if this were known. Zero-sum models, in which all energy contributions add to zero, show promise as a basic element of that design; the gravitational binding energy of the universe would then equal its mass energy. (*Mass energy* here includes rest-mass energy, electromagnetic energy, and the kinetic energy of the expansion of the universe.)

As the universe expands, its gravitational potential energy increases at the expense of its kinetic energy. At maximal expansion,[†] this kinetic energy would be zero. The stored rest-mass energy would then balance the gravitational potential energy. (The electrostatic and electromagnetic elements are assumed to balance themselves.) This leads to the interesting conclusion that, for a fully expanded universe,

$$\sum_i m_i c^2 = \sum_{i,j} \frac{G m_i m_j}{r_{ij}}$$
(3-12a),

$$\text{or} \quad c^2 \sum_i m_i = G \sum_i \left(m_i \sum_j \frac{m_j}{r_{ij}} \right)$$
(3-12b).

---

* The broadly accepted Lin-Shu density-wave theory, developed in the 1960s at MIT, maintains two well defined spiral arms but fails to account for the branched structures of many spiral galaxies. And although it provides a mechanism for maintaining simple spiral arms over time, it fails to explain their origin.[22] In this respect the mechanism proposed here may complement the Lin-Shu theory.

† One consequence of inflation is that the curvature of the universe should be zero. The 'maximum' expansion would then be an asymptotic limit.

Thus

$$G = c^2 \sum_j \frac{r_{ij}}{m_j} = c^2 \frac{r_{\text{eff}}}{M_U} \tag{3-13},$$

where r_{eff} is the effective radius of the fully expanded universe, and M_U is the mass of the universe.

An attempt to estimate M_U can be made by assuming that Equations (3-12) are *approximately* valid at present, that the expansion of the universe has slowed to the point that we're near the asymptotic limit. It then follows that the effective radius of the universe is the product of its age and the speed of light, some twelve billion (12×10^9) light-years. Given the gravitational constant, $G = 6.67 \times 10^{-8}$ dyne·cm²/gm, the total mass M_U of the universe should be roughly 10^{80} proton masses, or 10^{83} electron masses.[23]

With some additional effort, a better estimate of M_U should be possible. This is an area of current research.*

---

* One possibility is to use Equation (1-10) to calculate the mass of a resonance of N wavelengths, or $2N = 4 \times 10^{21}$ segments. This results in a universe of mass $M_U = (2N)^5 m_e \approx 10^{108} m_e$. However, this grossly overestimates the mass because it assumes a full electronic charge of e rather than the zero charge likely for the primordial entity. With some additional effort, a better estimate of M_U is possible.

The electron's unit charge derives from some 2×10^{21} internal twists. The first daughter resonances born of the primordial entity presumably had but a single twist, equivalent to a charge of $(1/N)e \approx 5 \times 10^{-22} e$. Generalizing $1/N$ to $(2N - n)/2N$ gives the correction factor to Equation (1-10) for all generations,

$$m_n = \frac{(2N - n) n^5 m_e}{2N} \tag{1-10'},$$

which approaches Equation (1-10) at low values of n. The first daughter resonances would thus have had a total mass of $(10^{108})(5 \times 10^{-22}) = 5 \times 10^{86} m_e$. Assuming conservation of energy, this would be the mass of the universe today.

The fact that this is three to four orders of magnitude greater than our earlier estimate suggests that the universe is not even close to its full expansion. It may well grow to the size needed for the results to agree. Taking M to be $5 \times 10^{86} m_e$ results in an r_{eff} of 40×10^{12} light-years—3000 times its current extent.

Another possibility under consideration is that mass may not be conserved between resonance generations. The masses of early resonances may have been quite small, and not reflect the total mass energy of the universe today. See Section 5-III for details.

3-VI. DARK MATTER

Based on a number of astronomical observations, a general consensus has formed that as much as 90% of the mass of the universe does not radiate sufficient light to be detected from Earth.[20] This mass is called *dark matter*.

Some of the most dramatic pieces of evidence for dark matter are observations of the rotational velocities of interstellar gas clouds in spiral galaxies. These clouds do not move in accordance with Kepler's Law.[20] This could only occur if the law of gravitation were to fail at galactic distances, or if there were massive distributions of unseen matter surrounding the galactic disks. Most cosmologists prefer the latter idea.

Kepler's Law arises from the balance between gravitation and the centrifugal effect on a planet of mass m and rotational velocity V, orbiting at radial distance r. That is,

$$\frac{GmM_r}{r^2} = \frac{mV^2}{r} \qquad\qquad (3\text{-}14\text{a}),$$

$$\text{or} \quad V = \left(\frac{GM_r}{r}\right)^{\!1/2} \qquad\qquad (3\text{-}14\text{b}),$$

where M_r is the total mass within radius r. In the case of our solar system, the dominant mass by far is the Sun. In this case M_r is nearly constant, and Equation (3-14b) can be rewritten to a very good approximation as

$$V \propto \sqrt{r} \qquad\qquad (3\text{-}15).$$

This is Kepler's Law. It predicts, for example, that the rotational velocity of Pluto is ten times less than that of Mercury, because Pluto is one hundred times farther from the sun.

Yet when astronomers observe the orbital motions of gas clouds in the disks of spiral galaxies, they find that the velocities are nearly constant with increasing distance from the center, once outside the bright inner hub.[20] The compelling conclusion is that, outside the massive galactic nuclei, M_r is not the near-constant one would expect. If V is independent of distance, then M_r must increase linearly with distance,

$$M_r \propto r \qquad\qquad (3\text{-}16).$$

The generally accepted interpretation of this result is that spiral galaxies are embedded in massive, spherically distributed halos of unobserved matter that extend well beyond their visible disks.[20]

Many candidates for this dark matter have been proposed, but none have met with broad acceptance. Massive dark objects, such as dead stars or black holes, have been all but ruled out by astronomical observations. Since dark objects of this nature can't be directly observed, astronomers have searched for the gravitational lensing they would cause. Some lensing has been observed, but not nearly enough to account for either the large amounts of dark matter or the distribution required by Equation (3-16).[24]

Recent investigations have focused on the possibility that neutrinos might have a very small rest mass and exist in sufficient numbers to account for much of the dark matter. Although neither the standard nor resonance models expect neutrinos to have mass (see Section 4-VIII), the possibility cannot be ruled out. The problem with identifying dark matter with neutrinos is finding a mechanism for the radial mass distribution indicated by Equation (3-16). Since neutrinos travel near to or at the speed of light, and have an extremely small interaction cross-section with ordinary matter, their distribution in the vicinity of a galaxy is expected to be nearly flat. Achieving the required radial distribution would require particles more massive than neutrinos, traveling at cooler velocities. The search has thus focused on what is called *cold dark matter*, possibly in the form of *weakly interacting massive particles*, or WIMPs.

The framework of the resonance model presented in Chapters 1 and 2 suggests its own candidate for dark matter. It is a heavy, neutral composite particle, a stable third-generation counterpart of the neutron, formed of D_s mesons and a tau lepton. This hypothetical particle has been coined the *psychron*, from the Greek ψυχρόν, 'cold one', in anticipation of its role. Besides its basic meaning of *cold*, the Greek ψυχρόν connotes *dull to the senses, lifeless, indifferent, insipid,* and *sluggish*—all fitting attributes for a basic constituent of cold dark matter. The suggested symbol is a capital Ψ.

Recall that the binding energies of the neutron tilt it only slightly from stability. It does not require a leap of faith to imagine that in the next generation of composite resonances, the balance of stability might favor a neutral particle.

Figure 3-2 shows a hypothetical structure of the psychron formed, by analogy with the neutron, from a hexagonal ring of alternating D_s^+ and D_s^- mesons encircling a core composed of a τ^-, a D_s^+, and a D^0 (taking D^0 to be the neutral partner of the D_s).

Just as the segmental annihilation (destructive interference) mechanism described in Chapter 2 and Appendix B affects the structure of the neutron, so should it affect the psychron. Figure 3-2 presents the most plausible configuration of the core. The reason that four of the five segments of the tau can be annihilated is that the amount of angular momentum involved, when transferred to the six remaining segments of the D_s mesons, does not exceed the fundamental limit of $\frac{1}{2}\hbar$.

The Ψ mass can also be estimated along the lines used for the neutron. The psychron's proposed constituents, eight D_s mesons and one tau lepton, gross 17.4 GeV. From this must come the binding energy, estimated in Chapter 2 to be 649 MeV per D_s, for 5.2 GeV; as well as the mass lost by destructive interference, $2\times(\frac{2}{3}m_{D_s} + \frac{2}{3}m_\tau) = 3.0$ GeV. This results in a heavy neutron-like particle massing some 9.2 GeV.

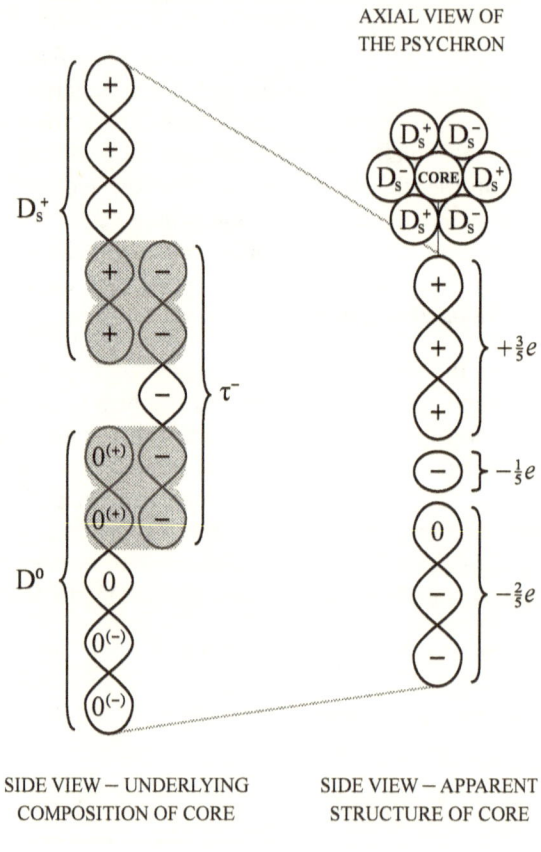

Figure 3-2. *THE PSYCHRON.*

The hypothetical psychron is the third-generation analogue of the neutron. (See Figure 2-6.) The shaded regions in the underlying composition, left, mark the segmental annihilations in the core of the particle. These produce the apparent structure of $\frac{1}{3}e$ fractional charges at right.

Such particles would exhibit the properties expected of cold dark matter. Because of their neutral charge and small size (D_s mesons and taus are quite a bit smaller than pions and muons), psychrons would have very little interaction with normal stellar matter. Due to their considerable mass, however, they would be gravitationally bound to galaxies. Those near a bright, dense galactic core would thermalize to stellar temperatures and be ejected. Eventually gravitation would slow this thermal motion and they would fall back into the galaxy. Over time, the psychrons would oscillate at some mean distance from the galactic center, resulting in a distribution with broad peaks at the turning points of the oscillations. This would match the distribution required by Equation (3-16) and produce the angular velocities observed in spiral galaxies.*

If these hypothetical particles do comprise a significant portion of the mass of the universe, then they should be observable through their predicted properties. Their neutral charge and low cross section present particular challenges to the experimentalist. However, nuclear magnetic resonance techniques should be able to detect the psychron's magnetic moment. This would be the -1.0 magneton of the tau, or $\mu_\tau(m_p/m_\tau) = -0.5 \, \mu_N$. A more careful reckoning, following the methods developed in Appendix C for the baryons, yields the smaller estimate $-0.1 \, \mu_N$.†

A hunt for psychrons would require a considerable volume of hard vacuum. One method of detection would be to impose a DC magnetic field and a tuned radio-frequency (RF) excitation throughout the vacuum. Any psychrons present would cause a small attenuation at the Larmor frequency, ω_L, of the psychron. RF pulse-echo measurements could be expected to enhance sensitivity. To eliminate residual background signals from normal

---

* To account for the velocities observed in Rubin's spiral galaxies, dark matter must extend well beyond the visible disks. Barred spirals, however, may be different. These galaxies exhibit an abrupt demarcation between an inner bar, which exhibits no wrapping, and the tightly wound outer arms. Conceivably, then, dark matter is confined to the bar region. The pendulum-like migrations imagined in the text could maintain such definite radial extents if the characteristic energies of the psychrons were determined by thermal excitation in the galactic cores. This would imply that spiral and barred-spiral galaxies differ in temperature, with the centers of barred spirals being substantially cooler. An extension of Rubin's work to barred-spiral galaxies would thus be of considerable interest as a first test of these ideas. If they are valid, one would expect angular velocities to vary linearly with radial distance within the bar regions, but to follow Kepler's law in the spiral arms.

† The psychron could have a substantial impact on our understanding of collapsed stars as well, including black holes. See the endnotes to this chapter for further discussion.

matter, and to obtain a large volume of vacuum, one might consider conducting this experiment to the lee of the space shuttle, where hard vacuums on the order of 10^{-14} torr are swept out by the craft.

3-VII. CONCLUSION

The concept of inertial mass arises naturally from the resonance model, and the equations of motion follow. This chapter presents a scenario for creation that may elucidate the big bang, inflation, galaxy formation, and the existence of stable subatomic particles.

The possible electromagnetic nature of gravitation, and *vice versa*, are plausibly handled within the framework of the resonance model. A number of consequences may be tested. For example, electronic charge should not be precisely conserved: the charge of the muon should be one part in 2×10^{21} less than that of the electron.

Finally, cold dark matter, which seems to make up most of the mass of the universe, may be at least partially composed of a stable neutron-like particle with a rest mass of some 9 GeV and a magnetic moment of about -0.1 nuclear magnetons.

Continued from p 73.

The general view of a black hole is that of a singularity within an *event horizon*. The event horizon is not a physical surface, but rather the distance from the singularity at which the escape velocity equals the speed of light. The distance $r \approx M_r G/c^2$, found by setting the orbital velocity, V, in Equation (3-14a) equal to the speed of light, c, turns out to be the exact solution for the event horizon of a maximally rotating black hole. Such an object is known as an *extreme Kerr black hole*.[25]

Since singularities can't be described mathematically, most physicists initially balked at the idea that they could actually exist. However, in the late 1960s Stephen Hawking and Roger Penrose demonstrated that, within Einstein's general theory, black holes always lead to singularities. They theorized that a singularity could be completely described with just three parameters: mass, angular momentum, and electronic charge. This is colloquially described by the adage 'black holes have no hair'—that is, that they have no internal structure.[26] This was assumed to follow from the fact that the very large average linear mass density required by the event horizon, $M_r/r = c^2/G$, exceeded the densest form of matter thought to be possible, the condensed neutrons predicted to form the cores of neutron stars.

Rather than sacrifice general relativity, the physics community reluctantly came to accept the idea of singularities. It was thought that once gravitation overwhelmed nucleonic matter, there would be nothing to prevent collapse to a geometric point. These conclusions are provisional, however (they assume that elementary particles are geometric points, for example), and singularities remain problematic. Stephan Hawking has said that, since the uncertainty principle is of paramount importance when dealing with singularities,

> general relativity brings about its own downfall by predicting singularities. In order to discuss the beginning of the universe [or the center of a black hole], we need a theory that combines general relativity with quantum mechanics.[27]

Psychrons, being both smaller and more massive than neutrons, could provide a partial solution to this conundrum by allowing a tractable internal structure for smaller black holes. Psychron stars could, in effect, be denser analogues of neutron stars, with a physical surface lying within the event horizon. More massive black holes, such as those believed to exist at the centers of most galaxies, would involve further collapse of the core, creating particles more massive than the psychron.[*] There would be any number of increasingly dense forms of matter, at ever higher particle generations, that would provide a series of stages at which stellar collapse would terminate no

---

[*] Or perhaps psychronic matter can be identified with the 'quark stars', denser than neutron stars but not quite dense enough to be black holes, that have recently been claimed by researchers with NASA's Chandra X-ray Observatory,[29] in which case all black holes would involve higher orders of matter.

matter how massive the star. Thus the resonance model may provide a mechanism for avoiding singularities while maintaining the full elegance of general relativity.

Intriguingly, recent observations of the accretion disk around a massive galactic black hole at the core of MCG-6-30-15, by ESA's XMM-Newton X-ray observatory, provides evidence for a strong magnetic field braking the hole's rotation.[28] The field appears to be too intense to be generated by the accretion disk, suggesting that it might emanate from the hole itself. Of course, no magnetic field could be produced by a geometric point, such as a singularity, but one would be possible if the center of the black hole were finite in size. Indeed, a core of high-order matter should generate just such a magnetic field through the self-alignment of the spins of its component particles, the same mechanism invoked for neutron stars. Outside the event horizon, any field would be gravitationally Doppler shifted to a zero-frequency (non-radiating) state, and would affect the surrounding accretion disk in ways similar to those observed at MCG-6-30-15.

[1] A.H. Guth, *The Inflationary Universe*
(Addison-Wesley, New York, 1997), 1-15.

[2] S.W. Hawking, *A Brief History of Time*
(Bantam Books, New York, 1988), 128-130.

[3] B. Parker, *Creation: The Story of the Origin and Evolution of the Universe*
(Plenum Press, New York, 1988), 185-195.

[4] J. Singh, *Great Ideas and Theories of Modern Cosmology*
(Dover, New York, 1961), 81-197;

E.J. Sternglass, *Before the Big Bang*
(Four Walls Eight Windows, New York, 1997).

[5] H.R. Pagels, *Perfect Symmetry*
(Bantam Books, New York, 1991), 347.

[6] G.D. Coughlan and J.E. Dodd, *The Ideas of Particle Physics*, 2nd ed.
(Cambridge University Press, 1991), 191-205.

[7] Parker, *ibid.*

[8] A.M. Hillas and T.E. Cranshaw, 'A comparison of the charges of the electron, proton and neutron.' *Nature* **184** (19 September 1959), 892-883.

[9] H. Bondi and R.A. Lyttleton, 'A comparison of charges on the electron, proton and neutron.' *Nature* **184** (19 September 1959), 974;

H.F. Dylla and J. King, 'Neutrality of molecules by a new method.'
Physical Review A, **7**, 4 (April 1973), 1224-1229;

M. Marinelli and G. Morpurgo, 'The electric neutrality of matter: a summary.'
Physics Letters B, **137** (1984), 439-442.

[10] M. Hamuy *et al.*, 'The absolute luminosities of the Calán/Tololo type Ia supernovae.' *The Astronomical Journal*, **112**, 6 (December 1996), 2391-2397.

[11] L.M. Lederman and D.N. Schramm, *From Quarks to the Cosmos*
(Scientific American Library, New York, 1989), 165-176.

[12] *Ibid.*, 172-181;

Guth, 185.

[13] P.J.E. Peebles, *Principles of Physical Cosmology*
(Princeton University Press, 1993), 408.

[14] Guth, 137-145, 283-284.

[15] M. Rees, *Before the Beginning*
(Perseus Books, Reading, Mass., 1997), 169.

[16] Peebles, 131-134.

---

[17] M. Begelman and M. Rees, *Gravity's Fatal Attraction* (Scientific American Library, New York, 1998), 77-84.

[18] M.J. Rees and J. Silk, in *Cosmology +1: Readings from Scientific American* (W.H. Freeman & Co., San Francisco, 1977), 53-62.

[19] H.C. Ferguson, R.E. Williams, and L.L. Cowie, 'Probing the faintest galaxies.' *Physics Today* (April, 1997), 24-30.

[20] V.C. Rubin, 'Dark matter in spiral galaxies.' *Scientific American* (June, 1983), 96-107.

[21] The Canada-France-Hawaii Telescope is managed by the National Research Council of Canada (CNRC), the Centre National de la Recherche Scientifique de France (CNRS), and the University of Hawaii.

[22] Time-Life Books, *Galaxies* (Time-Life, Alexandria, Virginia, 1988), 60-68.

[23] Lederman and Schramm, 175.

[24] Rees, 106-107;

Begelman and Rees, 85-95.

[25] E.F. Taylor and J.A. Wheeler, *Exploring Black Holes: Introduction to General Relativity* (Addison Wesley Longman, 2000), F1-13.

[26] Begelman and Rees, 13.

[27] Taylor and Wheeler, quote on F3.

[28] See the ESA information release of August 2001, and further links, at *http://www.esa.int*.

[29] NASA news release 02-082, 10 April 2002. Available online at *http://www.msfc.nasa.gov/news*.

Chapter 4

THE ELECTROMAGNETIC FIELD

The electromagnetic field is described with a notation that avoids any presumption of an inertial reference frame. This is useful for visualizing the formation of the standing-wave resonances identified with massive particles. The notation reveals the electromagnetic nature of neutrinos and resolves the violation of parity conservation in weak interactions. Finally, it clarifies the de Broglie waves that arise in conjunction with moving particles.

4-I. INTRODUCTION

A propagating electromagnetic field in free space, E, is conventionally represented with a wave notation,

$$E = A e^{i(\mathbf{k} \cdot \mathbf{r} - \omega t)} \tag{4-1},$$

where $\mathbf{k}$ is the wave vector and ω is the frequency.[1] The use of $\mathbf{k}$, ω, and the spatial and temporal coordinates $\mathbf{r}$ and t implicitly assumes the existence of an inertial reference frame. This monograph was inspired by an attempt to develop a representation of the electromagnetic field that does not require the *a priori* existence of such a coordinate system.

This objective was realized by regarding circularly polarized traveling waves in the electromagnetic field as being composed of positively and negatively charged units. Both are required for free propagation. Circular polarization predicates that the units follow helical trajectories; these are separated by a phase shift of π about their common axis of propagation, with a transverse electromagnetic field between them. In a vacuum this free propagation is at the speed of light. The charge units are polarized as either left-handed, L, or right-handed, R. (L and R are each defined to have an angular momentum of $\frac{1}{2}\hbar$.)

A right-handed traveling wave would be expressed as $|R^+, R^-\}$. The brace, $|\ \}$, indicates the direction of motion, to the right in this case, and the comma indicates the π phase shift between the two charge units. Rotational frequency and charge spacing are not explicit, since these parameters vary with the choice of coordinate system. Similarly, a wave traveling in the opposite direction with left-handed polarization is expressed by $\{L^+, L^-|$. Linearly polarized waves can be represented by superposing half waves with opposite polarization, such as $|½L^+ + ½R^+, ½R^- + ½L^-\}$. Here the electromagnetic field between the two counter-rotating waves lies in a plane whose orientation depends on their relative phase shift.

These four basic wave units, R^+, R^-, L^+, and L^-, each represent one-half unit of spin and charge. They should not be viewed as spatially localized, but rather distributed so as to produce the observed electromagnetic field. In any freely traveling wave they will exist in pairs, with zero net charge, moving collinearly at the speed of light. A standing wave can be formed by two superposed and oppositely directed traveling waves. The standing wave $\{L^+, L^- | R^+, R^-\}$, for example, is circularly polarized, its left- and right-hand components overlapping to create oscillating peaks between fixed nodes.*

The brace notation was intended merely as a conceptual means to visualize the electromagnetic field. However, it soon became clear that makes for a useful heuristic as well. The nature of an unknown element in a particle reaction may be determined by balancing charge and polarization in the brace notation. In Section 4-IV this is done to reveal the nature of neutrinos.

The new notation, with only two defined directions of propagation, may at first appear to describe just two possible orientations of an interaction—head-on collision or collinear propagation—whereas the propagation constant **k** in Equation (4-1) may assume any direction. However, if one considers two arbitrarily directed and overlapping electromagnetic waves, without reference to an inertial reference frame, these are really the only two possibilities. Either the waves propagate collinearly, or there is some coordinate frame in which they converge head on.

Another convenience of the brace notation is that one can immediately see when an electromagnetic field component satisfies the conditions for free-space propagation. Freely traveling waves are neutral and require full units of both positive and negative charge, such as are found in the linearly polarized wave $|½R^+ + ½L^+, ½R^- + ½L^-\}$. This is equivalent to saying that the

---

* To see this, let you left hand represent L, and your right hand R. Point your thumbs in the appropriate directions of motion (left to the left and right to the right in this case) and curl your fingers. Your fingers curl in the same direction, reinforcing each other.

symbols (+) and (−) correspond to half the fundamental unit of charge, $\pm\frac{1}{2}e$. Section 1-II and Appendix B show that fractionally charged components, such as $|\frac{1}{X}R^+, \frac{1}{X}R^-\}$, where $X > 1$ and $e' < e$, have phase velocities greater than c. Hence their free-space propagation is not possible, and they remain bound within massive (inertial) particles.

Significantly, such components can express linear and angular momenta. This is demonstrated in Section 4-IV, where the recoil of the muon created in pion decay is equated with a fractional electromagnetic-field component that has all the characteristics of a de Broglie matter wave.

4-II. A SIMPLIFIED ELECTRON

The brace notation will now be used to describe the particles of the first generation, the electron and positron. The simplified approach introduced in this section is incomplete in that it does not properly address the spin of the electron. Nonetheless, it provides both familiarity in using the notation and insight into the nature of the electron. The issue of the electron's spin is resolved in Section 4-VI.

In the simple standing wave represented by $\{L^+, L^- | R^+, R^-\}$, the electromagnetic fields components are antiparallel, canceling out at nodes every half wavelength along the axis of superposition. In the segments between these nodes the electromagnetic field varies, being greatest at the midpoints. The components where the transverse electromagnetic field originates, $\{L^+|$ and $|R^+\}$, are defined as *positive*, while those where the field terminates, $\{L^-|$ and $|R^-\}$, are defined as *negative*. When averaged over time, the rotating arcs of this field sweep out superimposed positive and negative volumes between the nodes, as shown in Figures 1-1. If one separates the positive and negative components of the standing wave, $\{L^+|R^+\}$ and $\{L^-|R^-\}$, one can see how charged particles might arise. The charges could separate if they were subjected to an external electric field, if they moved through a transverse magnetic field, or if they possessed sufficient kinetic energy to overcome their mutual attraction.

Electron-positron pair production may favorably occur when the standing electromagnetic wave described in Chapter 1 reaches or exceeds a frequency of ω_0 in its own inertial reference frame. This frequency corresponds to an energy of 1.02 MeV, the sum of the 0.51-MeV rest masses of the electron and positron.[2] Using the brace notation, pair creation and annihilation can be represented as

$$\{L^+, L^- | + | R^+, R^-\} \quad \leftrightarrow \quad \overrightarrow{\{L^-(\frac{1}{2})R^-\}} + \overrightarrow{\{L^+(\frac{1}{2})R^+\}}$$

$$\gamma \qquad\qquad \gamma \qquad\qquad\qquad e^- \qquad\qquad\qquad e^+ \qquad\qquad\qquad (4\text{-}2).$$

Here the factors of (½) signify half-wavelength (i.e., lowest-order) resonances, identified with the electron in Chapter 1. The electromagnetic units L$^+$, L$^-$, R$^+$, and R$^-$ are rearranged but conserved in both magnitude and direction. For clarity, the direction of spin is indicated by an arrow above the particle. These arrows are optional; their direction can always be determined by the right-hand rule:

Within a standing-wave resonance, the polarization is accompanied by a localized net circulation or spin—or lack of spin, as the case may be. The right side of Equation (4-2) establishes that the particles have a basis, L plus R, which produces a standing wave with an axial spin vector. This spin is defined to be collinear with |R} and anticollinear with {L|. Outside the standing-wave, the concept of net spin looses significance because there are no nodes to define a localized region. With a traveling wave it is preferable to think in terms of left- and right-handed polarization.

Looking at a spinning standing-wave resonance, such as {L$^+$(½) R$^+$}, from one direction along its axis, an observer sees a clockwise circulation, (L). When viewed from the opposite direction, the observer sees a counterclockwise circulation, (R). In contrast, the standing wave formed from the superposition of two oppositely spinning subcomponents, such as {½R$^+$(½) ½L$^+$} plus {½L$^+$(½) ½R$^+$}, appears identical regardless of orientation because the observer always sees (½L + ½R). This indicates a lack of spin or, equivalently, identifies the standing wave as linearly polarized—that is, as a boson. Although no such resonance occurs in the first generation alongside the electron, they do occur in all subsequent generations. The pi mesons discussed in Chapter 1 are one example.

The brace notation used in Equation (4-2) to characterize the electron and positron helps in understanding the stability of these particles. Clearly, neither resonance by itself can decompose. Neither contains both the positive and negative units necessary to produce traveling electromagnetic waves, which is the only decay mode available to particles of the first generation. The resonances have all the characteristics one would expect of an electron and positron, including an inherent uncertainty in the position of their electronic charge due to their wave nature.

If an electron and positron approach one another, they may orbit each other under the attraction of their opposite charges, forming an atom-like structure known as positronium.[2] This will radiate low-frequency electro-magnetic energy until it reaches its ground-state, the standing wave intro-duced in Figure 1-1(c). There is then a strong probability that the field will collapse into two divergent traveling waves, the reverse of Equation (4-2), with the frequency of each wave corresponding to the electronic rest mass of 0.511 MeV in the center-of-momentum reference frame.

Particle physicists refer to this as *annihilation*. They say that the electron and positron annihilate each other and that their mass energy is carried away by two gamma rays, or photons.

There is something not quite right in the association of $\{L^-(\frac{1}{2})R^-\}$ with the spin-$\frac{1}{2}$ electron. The subcomponents L^- and R^- each contribute one half to the unitary spins of the traveling waves $|L^-,L^+\}$ and $|R^-,R^+\}$. Thus the $\{L^-(\frac{1}{2})R^-\}$ resonance would also have a spin of one. This discrepancy will be resolved in Section 4-V with the benefit of the quantitative results developed in the next two sections. There the treatment of spin becomes more evident because the second generation is populated by both spinning and non-spinning particles, the muons and pions.

4-III. THE SECOND GENERATION: MUONS AND PIONS

The second-generation of the lepton family, the muons, are provisionally represented by $\{L^+(\frac{3}{2})R^+\}$ and $\{L^-(\frac{3}{2})R^-\}$, where the factor ($\frac{3}{2}$) signifies the three-half wavelength resonance illustrated in Figure 1-2. Muon formation is similar to the electron-positron pair production described by Equation (4-2), although a higher resonant frequency is required to generate the more massive muons. However, this description poses a problem, for the resonances exhibit the same unitary spins as the simplified electron.

Now, linear pions could emerge from two linearly polarized traveling waves through a process such as

$$\gamma_{\text{linear}} + \gamma_{\text{linear}} \tag{4-3a},$$

$$= \{\tfrac{1}{2}L^+ + \tfrac{1}{2}R^+, \tfrac{1}{2}L^- + \tfrac{1}{2}R^-| + |\tfrac{1}{2}L^+ + \tfrac{1}{2}R^+, \tfrac{1}{2}L^- + \tfrac{1}{2}R^-\}$$

$$= \{\tfrac{1}{2}L^+ + \tfrac{1}{2}R^+(\tfrac{3}{2})\,\tfrac{1}{2}L^+ + \tfrac{1}{2}R^+\} + \{\tfrac{1}{2}L^- + \tfrac{1}{2}R^-(\tfrac{3}{2})\,\tfrac{1}{2}L^- + \tfrac{1}{2}R^-\}$$

$$= \tfrac{1}{2}[\{\overleftrightarrow{L^+(\tfrac{3}{2})R^+}\} + \{\overleftrightarrow{R^+(\tfrac{3}{2})L^+}\}] + \tfrac{1}{2}[\{\overleftrightarrow{L^-(\tfrac{3}{2})R^-}\} + \{\overleftrightarrow{R^-(\tfrac{3}{2})L^-}\}]$$

which is equivalent to

$$\gamma + \gamma \;=\; \pi^+ + \pi^- \;=\; \tfrac{1}{2}(\mu^{+\uparrow} + \mu^{+\downarrow}) + \tfrac{1}{2}(\mu^{-\uparrow} + \mu^{-\downarrow}) \tag{4-3b}.$$

Here the structure of the pion may be visualized as two counter-rotating (opposite-spin) half-muons superimposed to form a single resonance,

$$\pi^+ \;=\; \tfrac{1}{2}\mu^{+\uparrow} + \tfrac{1}{2}\mu^{+\downarrow} \tag{4-4}.$$

This is the structure proposed for the pion in Section 1-III. Similarly, the π^0 may be conceived of as a half π^+ and a half π^- that share a common axis, and which are superimposed in spatial and temporal quadrature so that their net charge and magnetic moment are zero. A spatial quadrature, or angle of $\frac{\pi}{2}$ between the $\frac{1}{2}\pi^+$ and $\frac{1}{2}\pi^-$ current directions, would only be possible when combining linearly polarized (spinless) particles like pions. Thus one would not expect to find a neutral spinning fermion like a μ^0.

This neutral pion structure is supported by arguments similar to those used for the charged pions. The four conceptual subentities $\frac{1}{4}\mu^{+\uparrow}$, $\frac{1}{4}\mu^{+\downarrow}$, $\frac{1}{4}\mu^{-\uparrow}$, $\frac{1}{4}\mu^{-\downarrow}$ are the simplest, if not the only, combination that yields both zero net charge and zero spin. The orthogonal orientation, or quadrature, of the oppositely charged components is consistent with the observation that the oppositely directed gamma rays emitted when a π^0 decays have linear polarizations that are orthogonally oriented.[3]

The π^0 can be expected to have a very brief lifetime, like the positronium described in the previous section, because it contains both the positive and the negative components necessary for spontaneous decomposition into free-space electromagnetic waves. Such radiative decay would be the natural and expected decay mode. Indeed, the π^0 lives only about 10^{-8} times as long as a charged pion (0.84×10^{-16} *vs* 2.6×10^{-8} seconds),[4] and is one of the few particles known to decay into pure electromagnetic radiation.

A surprising source of encouragement is found in the branching ratios of this decay. 98.8% of the time the π^0 does decay into two gamma rays, but 1.2% of the time a single gamma ray is produced along with an electron-positron pair.[4] Heretofore there has been no explanation for these numbers. However, if the proposed π^0 structure is correct, they should be predictable from the geometry of the resonance.

This seems to be the case. Each of the four $\frac{1}{4}\mu$ subcomponents of the π^0 is equivalent, in turn, to two of the basic electromagnetic units L^+, L^-, R^+, R^-. Thus one can conceive of eight abstract units circulating within the π^0, each presumably spending one third of its time, on average, in each of the three segments. The probability is $(\frac{1}{3})^4 = \frac{1}{81}$, or 1.2%, that four of these units would exist simultaneously in the central segment of the π^0, where they could generate an electron-positron pair *per* Equation (4-2).*

The π^0 is an excellent subject for testing the various aspects of the resonance model because electrostatic effects, which would otherwise complicate the analysis, average to zero. Precise data should be able to establish whether the 1.2% branching ratio does in fact correspond to a value of $\frac{1}{81}$.

---

* The electron-positron pair would presumably need to originate from the symmetrical central segment of the pion, rather than one of the end segments, in order to match the electromagnetic boundary conditions of the decay.

4-IV. NEUTRINOS

The foundation has now been laid to explore the nature of the two well known neutrino families, the electron neutrino and antineutrino, $v_e \bar{v}_e$, and the muon neutrino and antineutrino, $v_\mu \bar{v}_\mu$. This section will focus on these four neutrinos. However, the resonance model can easily be extended to the tau neutrino family, $v_\tau \bar{v}_\tau$, which has been proposed for some time[5,6] but only recently observed.[7]

There are several sets of units that form mathematically null resonances. These will prove useful in the analysis of the neutrinos. They consist of components that cancel out at all spatial locations:

$$\{L^+| + |L^-\} = \{L^+|L^-\} = \{\tfrac{1}{2}L^+|\tfrac{1}{2}L^-\} = 0 \qquad (4\text{-}5a),$$

$$\{R^+| + |R^-\} = \{R^+|R^-\} = \{\tfrac{1}{2}R^+|\tfrac{1}{2}R^-\} = 0 \qquad (4\text{-}5b).$$

We can now quantify the decay of a pion into a muon, starting with Equation (4-4) in its brace-notation form:

$$\pi^+ = \{\overrightarrow{\tfrac{1}{2}L^+(\tfrac{3}{2})\,\tfrac{1}{2}R^+}\} + \{\overleftarrow{\tfrac{1}{2}R^+(\tfrac{3}{2})\,\tfrac{1}{2}L^+}\} \qquad (4\text{-}6a).$$

In order to form a muon, one of the components of the pion must flip, reversing its spin to align with the other. Mathematically, one may add an antiparticle, $\{\tfrac{1}{2}L^-(\tfrac{3}{2})\,\tfrac{1}{2}R^-\}$, to cancel the first term on the right, and then add a particle with opposite spin, $\{\tfrac{1}{2}R^+(\tfrac{3}{2})\,\tfrac{1}{2}L^+\}$, to effect the flip.

To balance the equation, these additions must be cast as null components:

$$\pi^+ = \{\overrightarrow{\tfrac{1}{2}L^+(\tfrac{3}{2})\,\tfrac{1}{2}R^+}\} + \{\overleftarrow{\tfrac{1}{2}R^+(\tfrac{3}{2})\,\tfrac{1}{2}L^+}\}$$

$$+ \underset{null}{\{\tfrac{1}{2}L^-|\tfrac{1}{2}L^+\}} + \underset{null}{\{\tfrac{1}{2}R^+|\tfrac{1}{2}R^-\}} + \underset{null}{\{\tfrac{1}{2}R^+|\tfrac{1}{2}R^-\}} + \underset{null}{\{\tfrac{1}{2}L^-|\tfrac{1}{2}L^+\}} \qquad (4\text{-}6b).$$

Rearranging the terms, while preserving both magnitude and direction,

$$\boxed{\pi^+ \rightarrow \underset{\mu^+}{\{\overleftrightarrow{R^+(\tfrac{3}{2})\,L^+}\}} + \underset{recoil\ of\ \mu^+}{|\tfrac{1}{2}R^+ + \tfrac{1}{2}L^+, R^-\}} + \underset{v_\mu}{\{\tfrac{1}{2}L^+ + \tfrac{1}{2}R^+, L^-|}} \qquad (4\text{-}6c).}$$

Equation (4-6c) is significant for several reasons. For one, it represents pion decay in purely electromagnetic terms, even though this decay is governed by the weak nuclear force. The first of the decay products on the right-hand side was tentatively identified with the muon in Section 4-III. However, this association is somewhat flawed in that the term exhibits a spin of one rather than the correct value of ½. The second and third terms are oppositely directed waves with the unusual characteristic of mixed circular and linear polarization (half spin ½ and half spin 0). One of these terms may represent the freely propagating mu neutrino, while the other may be bound to the muon as the momentum of its recoil. However, determining which is which is not possible at this point. In fact, if parity were conserved in this interaction, as expected from Maxwell's equations,[8] it would follow that the mu neutrino would propagate with equal probability to the left or right.

Experimentally, it is found that parity violation is characteristic of weak interactions. (See Appendix D.) The mu neutrino always propagates in a direction parallel to the magnetic moment of the muon. The rightmost term in Equation (4-6c) satisfies this condition. To be consistent with the ½ spin of the muon, the recoil must be the second term on the right, which has a spin of ½ counter to the unitary spin of the first term. Thus taken together they constitute a proper spin-½ muon. The wavelike nature of the recoil also corresponds to the de Broglie matter wave that characterizes a muon in motion. The linear component of the recoil momentum, $|½L^+, ½R^-\}$, is of course zero in the muon's inertial frame. But the spin component, $|½R^+, ½R^-\}$, is absolute. This resolves the discrepancy of the muon's spin.

The mu neutrino can be independently identified as the rightmost term in Equation (4-6c) through the concept, presented in Section 3-III, that electronic charge derives from a high-pitched internal twist. The positive and negative signs in the brace notation can thus be replaced with superscript 'r' and 'ℓ' for *right-handed* and *left-handed* internal twist,

$$\nu_\mu = |L^-, ½R^+ + ½L^+\} = |L^\ell, ½R^r + ½L^r\} \tag{4-7a}$$

One can see that the emission of a neutrino with left-handed polarization, traveling parallel to the magnetic moment of the muon, would add one half turn of right-handed pitch to the positive charge of the original pion. This would give the nascent muon its more fully developed electronic charge.

Maxwell's equations entail parity conservation as a consequence of their implicit assumption that electronic charge is conserved. This is obviously an extremely close approximation, but may not be absolutely correct. It is estimated in Chapter 3 that the difference in charge between the muon and electron is one part in 2×10^{21}. This is sufficient to explain the parity violation observed in neutrino emissions: The emission of a neutrino in the same direction as above but with right-handed polarization would subtract charge from the pion. This would move the decay product to a higher order, for example to a tau lepton, requiring an additional input of energy. Such an endothermic process could not occur spontaneously, and has not been observed. Thus parity violation is a consequence of energy conservation.

The muon can now be envisioned as two collinear units, $\{\frac{1}{2}R^+(\frac{3}{2})\frac{1}{2}L^+\}$, which precess about one another with the angular momentum of the recoil. Each subcomponent provides a spin of $\frac{1}{2}$, while the precession subtracts the same, for the correct net spin of $\frac{1}{2}$. The same applies to all leptons. The electron is described more thoroughly in this light in Section 4-VI.

Returning to Equation (4-6c), conservation of momentum requires that the neutrino and muon have equal but oppositely directed linear and angular momenta in the center-of-momentum coordinate system of the pion. Furthermore, the linearly polarized terms, $\{\frac{1}{2}R^+, \frac{1}{2}L^-|$ and $|\frac{1}{2}L^+, \frac{1}{2}R^-\}$, and the circularly polarized terms, $\{\frac{1}{2}L^+, \frac{1}{2}L^-|$ and $|\frac{1}{2}R^+, \frac{1}{2}R^-\}$, must have the same wavelength in the center-of-momentum system to satisfy the electromagnetic boundary conditions of pion decay.

This approach allows one to predict, almost by inspection, that the muon antineutrino emitted in the decay of a π^- meson would be

$$\bar{\nu}_\mu \ = \ |R^+, \frac{1}{2}L^- + \frac{1}{2}R^-\} \ = \ |R^r, \frac{1}{2}L^\ell + \frac{1}{2}R^\ell\} \tag{4-7b},$$

and that its propagation would again be parallel to the magnetic moment of the muon. This is precisely what is observed.

Similarly, the electron neutrinos can be identified as

$$\bar{\nu}_e \ = \ |R^-, \frac{1}{2}L^+ + \frac{1}{2}R^+\} \ = \ |R^\ell, \frac{1}{2}L^r + \frac{1}{2}R^r\} \tag{4-7c}$$

$$\nu_e \ = \ |L^+, \frac{1}{2}L^- + \frac{1}{2}R^-\} \ = \ |L^r, \frac{1}{2}L^\ell + \frac{1}{2}R^\ell\} \tag{4-7d},$$

based on reactions such as the decay of a muon into an electron,

$$\mu^- \ \rightarrow \ e^- + \bar{\nu}_e + \nu_\mu \tag{4-8}.$$

Here the notation correctly posits that the two neutrinos should be emitted parallel to the magnetic moment of the electron, which is itself parallel to the moment of the original muon.[9]*

In muon decay, both the $\bar{\nu}_e$ and the ν_μ carry away one-half unit of oppositely directed spin, for a resultant electron spin equal to that of the muon. Furthermore, since all four neutrinos identified in Equations (4-7) have full

---

* Incidentally, the fact that the muon and electron neutrinos are consistently left-handed has been considered something of a enigma. It prompted Nickolas Solomey to choose the title *'Where Have All the Right-Handed Neutrinos Gone?'* for a chapter in his book.[10]

The answer to his question of course is that, in the standard model, right-handed neutrinos have been called antineutrinos.

units of both positive and negative charge, they should propagate freely like photons. This is equivalent to saying they have zero rest mass.* And like photons, neutrinos should not be considered particles, since that term is reserved for massive (standing-wave) resonances in the resonance model.

The mu neutrino in Equation (4-7a) may be alternately expressed as

$$\nu_\mu \ = \ |(\tfrac{1}{2}L^-, \tfrac{1}{2}R^+) + (\tfrac{1}{2}L^-, \tfrac{1}{2}L^+)\} \tag{4-9}.$$

This form reveals that the ν_μ is equivalent to the sum of two photon-like electromagnetic-field components. The first component has half the terms of a linearly polarized photon, while the second is equivalent to half of a circularly polarized photon. (See Section 4-I.) Since they each contain only half the electronic charge, these abstract components cannot exist independently. They must remain linked intimately together throughout all space to satisfy the unitary charge requirements of free-space electromagnetic field propagation.

Interestingly, when the muon was discovered it was expected to have a second electron decay mode, with the emission of a gamma ray. Failure to find this radiation led to the proposal that neutrinos come in two varieties, one associated with the muon and one with the electron.[11] The treatment of muon decay in the resonance model brings the matter full circle, with the neutrino *being* the expected electromagnetic radiation, albeit with a mixed polarization that made it undetectable at the time.

The authors originally preferred the form of the notation in Equation (4-7a) to that in Equation (4-9), because the former clearly shows that the linearly and circularly polarized components of the neutrino may be phase locked, in this case with L^+ as the common phase reference. However, it was subsequently realized that this phase relationship may change as the neutrino propagates. This possibility is discussed in Section 4-VIII in relation to neutrino oscillations.

4-V. COLOR CHARGE IN QUANTUM CHROMODYNAMICS

High-energy electron-positron collision studies in the 1960s discovered that the production of charged pions exceeds that of muons by a factor of three.[12] There are, in effect, three pathways to generating a charged pion for each pathway to a muon. This unexpected finding led to the *ad hoc* concept

---

* 'Zero' rest mass is shorthand for saying that none of the mass-energy comes from rest mass. In other words, an object with zero rest mass cannot exist at rest, but always travels at light speed.

of *color charge*—a triplet with the names *red*, *blue*, and *green*—which forms the foundation of quantum chromodynamics (QCD). The three pathways for pion formation are represented in QCD by the two-quark color combinations of red and antired, blue and antiblue, and green and antigreen, all of which add up to a similar 'white' or 'colorless' pion.

Although QCD has proven useful, it provides no insight into the nature of color charge. Rather, color adds a new level of complexity to the subatomic domain. The resonance model, in contrast, provides an intuitive under-standing of the threefold yield of pions over muons.

According to Equation (4-6a), a charged pion can be described as two sub-components that are essentially half-muons with opposite spin. The muon, in turn, can be described by two parallel spin-½ subcomponents that counter-rotate relative to each other with an angular momentum of ½. Thus each subcomponent appears as a fractional (¼) muon with a fractional (±⅛) spin. Four subcomponents can take the form of either a pion or a muon, depending on their spin orientations.

Since their only distinguishing feature is their spin, conventionally represented as *up* or *down*, there are only two combinations that yield a spin-½ muon:

Muon States (net spin $\pm\frac{1}{2}\hbar$).

1. ↑↑↑↑
2. ↓↓↓↓

However, these same factors can combine in six different ways to form a spinless pion:

Pion States (net spin 0).

1. ↑↑↓↓   3. ↑↓↓↑   5. ↑↓↑↓
2. ↓↓↑↑   4. ↓↑↑↓   6. ↓↑↓↑

This 6 : 2 ratio of spin probabilities provides an intuitive understanding of the 3 : 1 production ratio of pions over muons, and thus a geometric explanation for color 'charge'.

4-VI. ELECTRON-POSITRON PAIR PRODUCTION

Section 4-II drew a simplified picture of electron-positron pair production that failed to account for their spin. With the insights gained in Sections 4-III and 4-IV, it is now possible to describe pair production more fully.

Our starting point is the formation of a standing wave from two oppositely directed, linearly polarized traveling waves. (Linear polarization was selected for this example to be consistent with the zero angular momentum of the $^1S_0$ ground state of positronium.[2])

$$\gamma_{\text{linear}} + \gamma_{\text{linear}} =$$
$$\{\tfrac{1}{2}R^+ + \tfrac{1}{2}L^+, \tfrac{1}{2}R^- + \tfrac{1}{2}L^-| + |\tfrac{1}{2}R^+ + \tfrac{1}{2}L^+, \tfrac{1}{2}R^- + \tfrac{1}{2}L^-\} \qquad (4\text{-}10a).$$

(Recall that commas represent the π phase shift between charge units.)

Combining like-signed terms,

$$\gamma_{\text{linear}} + \gamma_{\text{linear}} =$$
$$\{\overleftrightarrow{\tfrac{1}{2}R^+|\tfrac{1}{2}L^+}\} + \{\overrightarrow{\tfrac{1}{2}L^+|\tfrac{1}{2}R^+}\} + \{\overleftarrow{\tfrac{1}{2}R^-|\tfrac{1}{2}L^-}\} + \{\overrightarrow{\tfrac{1}{2}L^-|\tfrac{1}{2}R^-}\} \qquad (4\text{-}10b).$$

For electron-positron pair production, one each of the positive and negative field components needs to flip spin in order to resonate as material particles. Assuming it is the $\{\tfrac{1}{2}L^+|\tfrac{1}{2}R^+\}$ and $\{\tfrac{1}{2}R^-|\tfrac{1}{2}L^-\}$ components which flip, as in Equations (4-6), then the flipped positive component gives rise to a potential neutrino, $\{\tfrac{1}{2}L^+ + \tfrac{1}{2}R^+, L^-|$, and a recoil, $|\tfrac{1}{2}R^+ + \tfrac{1}{2}L^+, R^-\}$. The flipped negative component gives rise to a similar potential neutrino, $|\tfrac{1}{2}L^- + \tfrac{1}{2}R^-, L^+\}$, and recoil, $\{\tfrac{1}{2}R^- + \tfrac{1}{2}L^-, R^+|$. The oppositely directed neutrino factors cancel out, as can be shown with the null factors provided in Equations (4-5). The net result is a positron formed of two co-aligned field components, $\{\tfrac{1}{2}R^+|\tfrac{1}{2}L^+\} + \{\tfrac{1}{2}R^+|\tfrac{1}{2}L^+\}$, precessing relative to one another with a net angular momentum of $\tfrac{1}{2}$; plus a similar electron of co-aligned field components $\{\tfrac{1}{2}L^-|\tfrac{1}{2}R^-\} + \{\tfrac{1}{2}L^-|\tfrac{1}{2}R^-\}$. The net spin of the two particles is zero, consistent with the boundary conditions of their $^1S_0$ ground state.

Figure 4-1 illustrates the two precessing field components that constitute the electron. Because the precession is equal to the circulation rate of either component, but works against them, two circulations are required for a point of the field to complete a single orbit of the electron. This may explain the electron's gyromagnetic ratio of approximately two.

The electron can be reresented by the alternate notation $\{\tfrac{1}{2}L^\ell|\tfrac{1}{2}R^\ell\}$ + $\{\tfrac{1}{2}L^\ell|\tfrac{1}{2}R^\ell\}$ + a relative angular momentum of $\tfrac{1}{2}$. An *equator* may be defined as the plane where the relative phase between the components is zero, and *poles* may be defined as the points where the phase difference is π. (Note that if the twists add as vectors, the schematic cylinder in Figure 4-1 will be realized as a rounded electron. *Cf.* Figure 1-1(c).) Now in the { | direction the total twist is $N+1$, due to the additive effect of the contributions ℓ and L, while in the | } direction the twist is $N-1$ because R negates one of the N left-handed twists, ℓ. Thus with each circulation of $\{\frac{1}{2}L^\ell|\frac{1}{2}R^\ell\} + \{\frac{1}{2}L^\ell|\frac{1}{2}R^\ell\}$, the 'equator' would move by a distance of one half twist in the { | direction. However, the relative angular momentum counters this effect, resulting in a stationary electron.

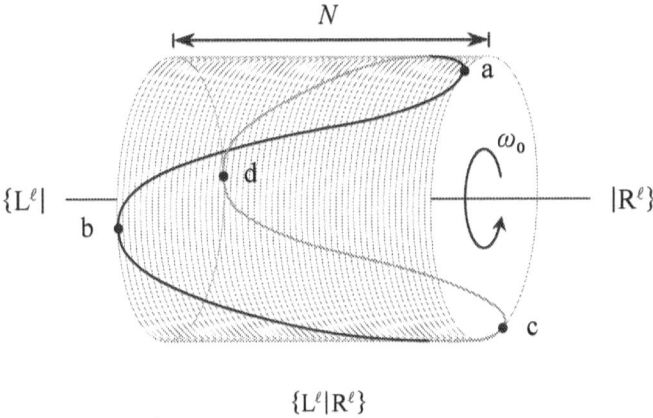

Figure 4-1. CONCEPTUAL FIELD REPRESENTATION OF THE ELECTRON.

The {L$^\ell$| and |R$^\ell$} units of the electron's resonant electromagnetic field are illustrated by a grey cylindrical envelope of N tight spiral twists. Since twist is left handed for negative charge, as described in Section 3-III, the spiral is tilted almost imperceptibly to the right. In a positron the spiral would be right-handed.

Due to the precession of the two units about each other, a conceptual point of the field at 'a', traversing unit {L$^\ell$|, will drift about the axis to 'b' by the time it reaches the other pole. It will then return along |R$^\ell$} to complete a single circulation. But by the time it does so, it will have been carried 180° around the axis of the electron to 'c'. It must therefore make a second circulation, through 'd', before it can return to its starting point.

Thus two circulations of the field are needed to complete a single orbit of the electron. This may be related to the electron's gyromagnetic ratio of approximately two.

To complete this picture, the trajectory of the conceptual point should follow the grey helical path N times around the axis in each direction. The pathway along the solid line is a "stroboscopic" image of the drift caused by the field's precession as it oscillates along the axis.

4-VII. DO NEUTRINOS HAVE REST MASS?

Ever since the discovery of neutrinos, there has been speculation that they may have some small but finite rest mass. The standard model favors zero rest mass. In the past, all experimental claims of a finite mass have proven inaccurate or unsupportable. Nevertheless, the prevailing sentiment, as expressed by Gordon Kane, is,

> The neutrino masses are not measured yet; they are consistent with zero, but most theorists do not expect them to be zero.[13]

While a first draft of this manuscript was in preparation, new experimental support for a neutrino rest mass was announced with considerable fanfare in the popular press. The June 5, 1998, issue of the *Los Angeles Times* carried the story on its front page. The work was conducted by an international team of scientists using a neutrino detector, called the Super Kamiokande, built one kilometer under Mt Ikena in Japan and containing fifty-thousand metric tons of ultra-pure water.

The excitement was caused by the finding that the flux of mu-neutrinos produced by cosmic rays striking the Earth's atmosphere above the detector was measurably different from the flux detected from the opposite side of the globe 13 000 km away. This had not been predicted, since neutrinos were expected to pass through the earth with no appreciable attenuation.

Since the Super Kamiokande only detects muon and electron neutrinos, the reduction in mu-neutrino flux could be explained if a fraction of these neutrinos somehow oscillated into another 'flavor', possibly tau neutrinos, during their 13 000-km journey through the Earth. The researchers believe such oscillations are only possible if the mu neutrino has a rest mass.*

In fact, there are other interpretations of the experimental results. One possibility is that the two subcomponents of the mu neutrino expressed in Equation (4-9) have subtly different phase velocities, so that the phase difference between the subcomponents oscillates with distance. This possibility becomes apparent when one sees Equation (4-9) in the form that associates electronic charge with the internal twist,

$$\nu_\mu = |(\tfrac{1}{2}L^\ell, \tfrac{1}{2}R^r) + (\tfrac{1}{2}L^\ell, \tfrac{1}{2}L^r)\} \qquad\qquad (4\text{-}9').$$

---

* Their conclusion is based on an analysis of neutrino mixing. If neutrinos oscillate between two flavors, then the characteristic mixing length, Λ, is $E/1.27\Delta(m^2)$, where E is the neutrino energy, in GeV; $\Delta(m^2)$ is the difference in the squares of the neutrino masses, in eV^2; and Λ is in km. Thus a finite mixing length implies a rest mass for one or both neutrino flavors. The best fit for $\Delta(m^2)$ is 0.0022 eV^2.[14]

The second term, ($\frac{1}{2}L^\ell$, $\frac{1}{2}L^r$), is equivalent to a half-photon with L^ℓ and L^r units. In L^r, the single rotation of the L opposes the large number of right-handed twists, r, associated with positive electronic charge. Conversely, the first term, ($\frac{1}{2}L^\ell$, $\frac{1}{2}R^r$), is additive in the effects of both L on ℓ and R on r. This difference would cause the relative phase velocities of the two components of a neutrino to vary by about one part in $N = 2 \times 10^{21}$.

If the components do differ, then the characteristic mixing length between constructive and destructive interference of the components will vary with neutrino energy. The mixing length in our example, 2×10^{21} wavelengths, is some 100 miles at a neutrino energy of 5 GeV. This is similar to the 400-mile estimate reported,[14] and could explain the results observed at the Super Kamiokande detector without assuming neutrino rest mass.

Note that in the resonance model's interpretation, the oscillation length of 2×10^{21} wavelengths varies *inversely* with neutrino energy. In contrast, if the oscillations are caused by neutrino rest mass, then the oscillation length will vary *directly* with energy.* This disagreement should soon be resolved.

This mechanism would also account for the 'missing' 70% of the solar neutrino flux.[15] Because the solar flux involves electron neutrinos which are substantially less energetic than cosmic rays, the mixing length would be greater than the diameter of the Earth. But it would be considerably less than the distance of the Earth from the Sun, allowing oscillation in transit.†

---

* Both models explain the observations at the Super Kamiokande. The observed neutrinos are born of cosmic rays striking the atmosphere. If neutrinos have mass, then the oscillation length, Λ, is directly proportional to neutrino energy. At low energies, Λ would be shorter than the distance between the detector and the point of origin, both for neutrinos born overhead and on the opposite side of the Earth. Thus the numbers of upward to downward neutrinos would be similar, as observed. As E increases, the oscillation length of the overhead neutrinos would eventually exceed the distance to the detector, and their number would exceed the number of neutrinos detected from below, again as observed.

In the resonance model, Λ is directly proportional to a fixed number of neutrino wavelengths. These are inversely proportional to their frequency, and thus to their energy. At low energies, the Λ of both upward and downward neutrinos would exceed the distance to the detector, and the numbers detected would be similar. As energy increases, the oscillation length of the neutrinos born on the opposite side of the Earth would eventually drop below this distance, and relatively more neutrinos would be detected from above. Again, this is what is observed.

† If this explains the solar neutrino problem, it raises the possibility that circularly polarized photons might exhibit a related mixing phenomenon. For visible light, $\lambda \approx 0.5$ μm, a mixing length of 2×10^{21} wavelengths would be approximately 0.1 light years. Such mixing may be observable in the spectra of nearby stars.

From the perspective of the resonance model, then, the claims of neutrino rest mass do not stand. All neutrino flavors are believed to be free-space electromagnetic waves. The only mechanism that can realize rest mass is the standing-wave resonance.

4-VIII. CONCLUSION

Both photons and neutrinos are identified as freely traveling modes of the electromagnetic field. Neither are considered 'particles' in the resonance model: Since mass is characterized as an electrodynamic effect of standing-wave resonances in Chapter 3, no traveling wave, photon or neutrino, is expected to have a rest mass.

TABLE I. *Summary of brace notation identities for traveling electromagnetic waves.*

PHOTONS

$\gamma_{\text{left-polarized}} \;=\; |L^+, L^-\}$ *or* $|L^r, L^\ell\}$

$\gamma_{\text{right-polarized}} \;=\; |R^+, R^-\}$ *or* $|R^r, R^\ell\}$

$\gamma_{\text{linear}} \;=\; |\tfrac{1}{2}R^+ + \tfrac{1}{2}L^+, \tfrac{1}{2}R^- + \tfrac{1}{2}L^-\}$

or $|\tfrac{1}{2}R^r + \tfrac{1}{2}L^r, \tfrac{1}{2}R^\ell + \tfrac{1}{2}L^\ell\}$

NEUTRINOS*

$\nu_e \;=\; |L^+, \tfrac{1}{2}R^- + \tfrac{1}{2}L^-\}$ *or* $|L^r, \tfrac{1}{2}R^\ell + \tfrac{1}{2}L^\ell\}$

$\bar{\nu}_e \;=\; |R^-, \tfrac{1}{2}L^+ + \tfrac{1}{2}R^+\}$ *or* $|R^\ell, \tfrac{1}{2}L^r + \tfrac{1}{2}R^r\}$

$\nu_\mu \;=\; |L^-, \tfrac{1}{2}R^+ + \tfrac{1}{2}L^+\}$ *or* $|L^\ell, \tfrac{1}{2}R^r + \tfrac{1}{2}L^r\}$

$\bar{\nu}_\mu \;=\; |R^+, \tfrac{1}{2}L^- + \tfrac{1}{2}R^-\}$ *or* $|R^r, \tfrac{1}{2}L^\ell + \tfrac{1}{2}R^\ell\}$

---

* Because the tau is believed to have one unit less twist than the muon, the tau neutrino and anti-neutrino may be similar to the ν_e and $\bar{\nu}_e$ but, tentatively, also with one unit less twist. That is:

$\nu_\tau \;=\; |L^{r-1}, \tfrac{1}{2}R^{\ell-1} + \tfrac{1}{2}L^{\ell-1}\}$

$\bar{\nu}_\tau \;=\; |R^{\ell-1}, \tfrac{1}{2}L^{r-1} + \tfrac{1}{2}R^{r-1}\}.$

A 'brace' notation has been developed to assist the visualization of both traveling and standing electromagnetic waves. Table I summarizes the brace notation identities of traveling waves. The capital R and L in these equations represent *right* and *left* circular polarization, while the superscript r and ℓ refer to the chirality of the twist associated with electronic charge. The absolute values of r and ℓ are believed to be very large, on the order of 2×10^{21}. (See Chapter 3.)

The brace notation provides further insight into the nature of matter by associating linear and angular momenta with electromagnetic field components, effectively de Broglie matter waves, bound to massive particles.

[1] P. Lorrain and D. Corson, *Electromagnetic Fields and Waves*, 2nd ed.
     (W.H. Freeman & Co., New York, 1962), 459-467.

[2] R.B. Leighton, *Principles of Modern Physics*
     (McGraw-Hill, New York, 1959), 624-626.

[3] Plano *et al.*, *Physical Review Letters*, **3** (1 December 1959), 525.

[4] Leighton, 636-637.

[5] G.D. Coughlan and J.E. Dodd, *The Ideas of Particle Physics*, 2nd ed.
     (Cambridge University Press, 1991), 161-162, 196-197.

[6] L. Lederman, *The God Particle*
     (Dell Publishing, New York, 1993), 324, 360-361.

[7] B. Schwarzschild, 'The tau neutrino has finally been seen.'
     Physics Today (October 2000), 17-19.

[8] P. Hartman, *Nuclear and High Energy Particle Physics*
     (Cornell University, Ithaca, NY, *c.* 1960), 72.

[9] Donald H. Perkins, *Introduction to High Energy Physics*
     (Addison-Wesley, Menlo Park, 1987), 222-223.

[10] Nickolas Solomey, *The Elusive Neutrino*
     (Scientific American Library, New York, 1997).

[11] Coughlan and Dodd, 75.

[12] L.M. Lederman and D.N. Schramm, *From Quarks to the Cosmos*
     (Scientific American Library, New York, 1989), 106-107, 151-155.

[13] G. Kane, *Modern Elementary Particle Physics*
     (Addison-Wesley, New York, 1994), 8.

[14] B. Schwarzschild, 'Cosmic ray showers provide strong evidence of neutrino
     flavor oscillation.' *Physics Today* (August 1998), 17-20.

[15] Solomey, 71-77.

Chapter 5

EPILOGUE

This monograph presents the case that massive subatomic particles are standing-wave resonances of the electro-magnetic field, while the massless photons and neutrinos are freely traveling electromagnetic wave modes. It affords insight into the natures of the strong, weak, gravitational, and electromagnetic forces, suggesting that they have a common origin. These concepts form the conceptual frame-work for a unification of the forces of nature that is quite different from the popular gauge theories of recent decades.

5-1. SUMMARY

Three premises form the basis of this monograph: (1) Electromagnetism and gravitation are united in five dimensions; (2) the electromagnetic field may form resonant standing waves that define an inertial system, in which massive particles may materialize; and (3) that the decay of a high-order resonance created our universe. These premises have profound implications:

1. Electromagnetism and general relativity can be united in a five-dimensional worldview.

There has recently been considerable renewed interest in the synthesis of electromagnetism and general relativity that was formulated by Theodor Kaluza in 1919 and refined by Oskar Klein several years later. Kaluza showed that by adding a mathematical fifth dimension, or degree of freedom, to Einstein's four-dimensional space-time continuum, Maxwell's equations for electromagnetism can be derived from general relativity.[1] By treating this fifth dimension as harmonic, with a period related to Planck's constant, Klein effected a transition to Schrödinger's quantum theory as well.[2] While Kaluza had not attributed any physical significance to this fifth dimension, introducing it solely to 'increase the number of components of the metric tensor', theorists have since associated it with a 'collapsed' spatial coordinate that may be the source of electronic charge.[3] Kaluza's dimension implied that the metric was cylindrical. Later, the dimension itself was characterized as circular: Peter Bergmann, one of Einstein's students, said 'it is assumed that those geodesics which connect a point with itself around the tube intersect themselves at an angle zero, in other words, that they are continuous, closed lines.'[4]*

Kaluza-Klein theory was the foundation of modern string theories, which include many more dimensions in an attempt to encompass the strong and weak forces and thus achieve a 'theory of everything'. These theories developed to the point where the number of dimensions grew to an almost intractable 26! However, with an assumption that fermions and bosons are associated in *supersymmetric pairs*, the number drops to ten in any of the five so-called *superstring* theories, and to eleven in a related field theory— all of which may be subsumed by the *M-theory* that is currently in vogue.[5] Still, this is a daunting task, even for the best mathematical physicists.

It seems that all string theories continue to make the assumption that strings form closed loops with circumferences equal to the Planck length. This is the main reason they are so far off with their mass calculations: even the lightest particle would be 16 orders of magnitude more massive than the proton.[6] This remains a huge problem.

In the resonance model the loops are thought to be much larger, comparable to the Compton wavelength. Furthermore, they are not perfectly closed, but form a helix that advances on the scale of the Planck length for each loop. This is the basis for the concept of twist described in Premise 3.

The original Kaluza-Klein theory may well be sufficient for a unified theory of physics. Its five dimensions (three spatial dimensions, x, y, z; a temporal dimension, t; and a collapsed dimension, e, that manifests itself through electronic charge) are precisely those five parameters whose signs may be reversed under charge-parity-time (CPT) symmetry without altering the nature of a phenomenon. That is, the mirror image $(-x, -y, -z)$ of any process, involving antiparticles $(-e)$ and reversed in time $(-t)$, is indistinguishable from the original.

The broadly accepted CPT theorem has not been offered before as support for Kaluza's metric. It suggests that five is indeed the correct number of dimensions for our universe.

---

* Klein demonstrated that the scale of this dimension would be on the order of the Planck length, $\approx 10^{-33}$ cm, and therefore not ordinarily perceptible.

2. Massive particles materialize from allowed standing-wave resonances of the electromagnetic field.

This novel premise is dubbed the *resonance model*. It answers Feynman's question, quoted in the Preface, *why is it that the mass of the muon compared with the electron is exactly 206?* The resonance model elucidates the masses of elementary particles, both known and postulated. The masses of compound particles, including the baryons, are calculated to an accuracy of about one percent. Furthermore, the heretofore elusive basis for the magnetic moments of the proton and neutron is resolved, and the values of the other baryons are calculated. Significantly, the estimated moment of the Σ^0 is negative, while the standard model predicts it to be positive. (See Appendix C for an analysis of the Σ^0 moment.) This will serve as a test of the resonance model, if even the sign of the moment of this very short-lived particle can be determined.

The family of allowed standing-wave resonances includes all half-integral numbers of wavelengths up to some maximum value, as developed in Premise 3 below. The electron is identified with the first order, a resonance of one half wavelength. The second order's three half-wavelength segments constitute the muons and pions. These can be imagined as very short strings of three pearls, each having a charge of $\frac{1}{3}e$. The segments trace out full volumes in the three-dimensional spinning muons, but are flat in the two-dimensional, spinless pions, clarifying the Pauli exclusion principle.

The analysis of pion structure has similarities to the concept of quarks in QCD. Specifically, the two identical end segments of the π^+ have the combined $+\frac{2}{3}e$ charge of an up quark, while the middle segment, which is distinct due to distortion by the end segments, has the apparent $+\frac{1}{3}e$ charge of an antidown quark. The resonance model, however, goes beyond the standard model in predicting that the next elementary particle is composed of five fractionally charged segments, each with an apparent charge of $\frac{1}{5}e$. This resonance is tentatively identified with the D_s meson. Other heavier particles are believed to exist with apparent internal charges of $\frac{1}{7}e$, $\frac{1}{9}e$, ..., $\frac{1}{2n+1}e$, where n is any integer up to some very large value.

The masses of elementary particles are derivable from the mass of the electron. For the next elementary particle after the massive *top*, calculations predict a mass of approximately 780 GeV. It will likely be observed early this century. Supporters of the standard model may be inclined to associate this particle, when discovered, with the elusive Higgs boson. However, the resonance model predicts that there is a nearly limitless series of increasingly heavy undiscovered particles, with characteristics similar to the Z^0.

Although many prior attempts have been made to explain matter in purely electromagnetic terms, a force stronger than electromagnetism seemed necessary to explain nuclear interactions. Through this second premise, the strong force is revealed to be fully electromagnetic in nature, enhanced by the superluminal phase velocities that attend fractional charges. Higher-order mesons, with charges that are more highly divided than in the pions, should effect even stronger forces.

Inertia and mass are epiphenomena of resonances in the electromagnetic field. Various permutations of polarization of the resonances correspond to distinct families of particles. Similarly, freely traveling waves include, in addition to the photon, several mixed polarization modes identified as neutrinos. This flies in the face of the standard model's conception of both the electron and its associated neutrino as being particles that are intimately related through a gauge transformation described by electroweak theory. The difference may be more of semantics than substance. However, the standing-wave description of massive particles leads to a deeper level of insight, specifically in explicating the masses and magnetic moments of elementary particles, accounting for the parity violation in weak decay, and in understanding the electromagnetic nature of the strong force.

3a. The series of standing-wave resonances ends with a single entity, consisting of some 2×10^{21} wavelengths but otherwise null. This was *the* primordial entity that began our universe.

This third premise is required to characterize the nature of electronic charge, gravitation, the weak force, and parity violation.

Care has been taken not to call the primordial entity a particle. It seemingly had no characteristics apart from number—no mass, no energy, no force, no space, and no time. It specifically did not have any electronic charge. In effect, this entity was a Fourier-like component of the original nothingness. Any value for the number of wavelengths would have been possible. It turns out that a value of approximately 2×10^{21} is associated with our universe.

3b. The entity decayed in a systematic cascade. Each generation during the decent inherited an outwardly diminished number of segments, but compensated for this by twisting internally, so that the total number of wavelengths in the five-dimensional metric remained a constant 2×10^{21} for every particle generated.

Although perhaps overly metaphorical, one may think of the twist as a definition of existence itself. As it increased in each generation, one could say that the particles were coming into existence.

The final particles of the cascade have a single, relatively large segment, but contain some 2×10^{21} tightly wound twists. Those with left-handed twist are known as electrons, the direction of twist defining negative electronic charge, while those with right-handed twist are positrons.

Every electron thus carries within it a remnant of the primordial resonance. These twists are tightly packed. Their mean separation, λ_C/N, is comparable to the Planck length, ℓ_{Pl}. Indeed, the reduced Compton wavelength, $\lambdabar_C = \hbar/m_e c = 3.9 \times 10^{-11}$ cm; the Planck length, $\ell_{Pl} = (G\hbar/c^3)^{1/2} = 1.62 \times 10^{-33}$ cm; the electromagnetic force; and gravitation are related to the number of twists in the electron, N, by

$$N^2 \equiv \frac{F_e}{F_g} = \alpha \left(\frac{\lambdabar_C}{\ell_{Pl}} \right)^2 \tag{5-1},$$

where α is the fine structure constant, $e^2/\hbar c$.

The constant N is formally defined in the next section.

Since α is dimensionless, we can define a new unit of length, called the *Planck wavelength*,

$$\lambdabar_{Pl} \equiv \frac{\ell_{Pl}}{\alpha^{1/2}} \tag{5-2}.$$

This takes on a particularly interesting form that combines four fundamental constants of nature,

$$\lambdabar_{Pl} = \frac{G^{1/2} \hbar}{c e} \tag{5-3}.$$

Equation (5-1) can be rewritten in terms of $\lambdabar_{Pl}$,

$$\lambdabar_C = N \lambdabar_{Pl} \tag{5-4}.*$$

That is, N Planck wavelengths make up the Compton wavelength of the electron. Thus the resonance model depicts the electron as having N twists along its axis, twists with circumferences comparable to the Compton wavelength and spaced a Planck wavelength apart, as shown in Figure 4-1.

This is reminiscent of Yōichirō Nambu's original hadronic string theory of 1970. Nambu's strings (not loops) were about as long as the Compton wavelength of a nucleon, and tied its quarks together. These strings needed a tension of some 15 tons to generate the resonant frequency of the proton.[7]

---

* Or $\lambda_C = N \lambda_{Pl}$. The symbol $\lambdabar$ stands for the 'reduced' wavelength, or radius, $\lambda/2\pi$, by analogy with $\hbar \equiv h/2\pi$.

Strings were revived in 1984 as a 'theory of everything'. However, in order to describe all forces, their scale was made much smaller—on the order of the Planck length, and with correspondingly higher tension—because it was thought that gravitation could only be unified with other forces at extremely high energies.[8] This leads to the mass problem mentioned earlier.

String theorists were happy with how Kaluza described his fifth dimension because it encompassed Maxwell's equations which, in turn, assumed the conservation of electronic charge. In the resonance model, on the other hand, twist is a function of particle generation, and electronic charge is not precisely conserved between generations. This adds a layer of complexity to an already difficult model. However, if this idea is correct, physics may be worked out in 5 dimensions, rather than in the 10 or 11 of string theory.

3c. The specific value of N, 2.04×10^{21}, comes from the assumption that a single twist should generate an electrostatic field equal to the gravitational field of a single segment of a resonance.

If this assumption is valid, then the square root of the ratio of the electrostatic and gravitational forces between an electron and positron, 2.04×10^{21}, defines the number of twists in the electron as well as the number of wavelengths in the primordial entity.

The decomposition of the primordial entity into a cascade of daughter resonances may have been the trigger for the big bang. Its expansion was driven by the release of kinetic energy, the amount of which increased with each successive generation.

At the moment of creation, it is speculated that the primordial null entity of 2×10^{21} wavelengths divided into two equal parts, each with one less segment.* These diverged at a high velocity.† As they separated, segment by segment, the first ticks of time began. Then they too decayed.

---

* It is presumed that the daughters had one-half of a left-handed twist each, on the scale of the compact primordial entity, and that they traveled in opposite directions so that their sum remained the null value of Equation (4-5a). Partition into right-handed resonances would have been equally probable, but the bias of electrons over positrons in our universe suggests that left-handedness is basic.

† The daughters had common axes with the primordial entity, but their spins were oppositely directed, analogous to the oppositely directed subcomponents of the pion. Since their segments were slightly larger than those of their parent, the only way for the daughters to originate at all was to be born with a velocity that would reduce their segments, through Lorentz contraction, to the size of the primordial segments in the center-of-momentum reference frame. In this way the daughter resonances would have satisfied the boundary conditions of their formation.

Succeeding generations down the decay chain furthered the transformation of the 2×10^{21} wavelengths of the initial nothingness into the growing internal twist (*i.e.*, charge) of the emerging quasi-particles.

The rate of expansion of the universe was driven by the phase velocities of the resonant cascade, and initially exceeded the speed of light by a factor of $(2 \times 10^{21})^2 = 4 \times 10^{42}$, slowing to c with the generation of electrons at the end of the inflationary period. This is consistent with the work of J. W. Peebles,[5] who established that extraordinary speeds are required to explain inflation—on the order of 10^{35} c when averaged over the inflationary era.*

Thus this third premise not only characterizes the inflationary period, but explains why the expansion slowed over time. It also suggests how an originally homogeneous universe broke up into the discrete galaxies we observe today, as well as to why spiral galaxies are typically substantially less massive than ellipticals.

Taken together, these premises form the framework for a unified theory of all physical phenomena that merits serious consideration.

5-II. A FRAMEWORK FOR UNIFICATION

Concepts of supersymmetry are being explored by string theorists, who can greatly simplify their analyses of bosons and fermions by pairing them together. They have not, however, been able to identify suitable supersymmetric partners for known particles that are consistent with the standard model. Indeed, they have seriously postulated the existence of such makeshift counterparts as the *photino* for the photon and the *selectron* for the electron.[10] These theorists are forced to assume that none of these exotic particles have been observed because they are exceptionally massive. In fact, because of these large masses, some suggest that these particles may be related to the cold dark matter that is presently believed to account for 90% or more of the mass of our universe.

Here the authors feel that the rigid framework of the standard model has driven theorists to grasp for solutions. The three premises summarized in Section 5-I establish an alternative framework that may shed some light on the difficulties encountered by string theorists. However, such speculation should soon give way to experiment. Steven Weinberg expects this to occur within twenty years:

---

* Inflation was devised by Alan Guth in 1981 to explain why the universe should be so nearly flat.[9] It has become broadly accepted, and there are hopes that it will soon be experimentally verified.

> All these ideas [about supersymmetric pairs] share another common feature: they require the existence of a zoo of new particles with masses not much larger than 1000 GeV. If there is any truth to these ideas, then these particles should be discovered before 2020 at the Large Hadron Collider, and some of them may even show up before then at Fermilab or CERN.[11]

If the resonance model has merit, then this anticipated zoo of elementary particles will never materialize. The only elementary particles predicted for this mass range have rest masses of approximately 780 and 1450 GeV (see Chapter 1), although composite particles may of course also exist.

Nevertheless, a limited amount of symmetry does emerge from the resonance model. The partner of the photon is the familiar neutrino, as described in Chapter 4. The electron has no partner, as explained in Chapter 1, but the muon and pion are paired—as may be other, heavier particles such as the D_s meson and τ lepton. The similar masses of the lepton-boson pairs, and their structural similarity in the resonance model, suggest that the symmetry may be unbroken (exact). The standard model, on the other hand, requires broken symmetry in order to account for the existence of mass.[12]

On a more speculative level, the authors argue in Chapter 3 that cold dark matter may be associated with a hypothetical neutron-like particle that is as stable as the proton. This would fit into the logical sequence of composite particles, and should have a rest mass of approximately 9.2 GeV. Gravitational attraction would cause swarms of these particles to form loose halos about large galactic masses, much like the presumed distribution of cold dark matter. Although the particle would be inert, its principal interaction with baryons being a relatively weak scattering, detection should be possible because it is predicted to have a measurable magnetic moment.

5-III. PERSPECTIVE

The stream of positive results which emerged from the second premise, the resonance model itself, inspired further work. This eventually led to the concept of twist and to the realization that, taken together, these form a solid basis for unifying the four forces observed in nature.

The foundations of this work were born of an attempt to describe traveling electromagnetic waves isolated from all material reference points. Such a description would be highly abstract, since no absolute frequency, wavelength, energy, direction, or physical extent can be defined without reference to an inertial frame.

This prompted the question of just what is required for an inertial frame. Is a material reference point needed? Or some inertial mass?

It seemed that a necessary and perhaps sufficient condition for an inertial reference frame was the existence of a standing wave, such as can result from the interaction of freely traveling waves. The nodes of the standing wave would establish a unit of measure, or metric, and the reference frame in which the nodes appeared stationary would be inertial.

It appeared that matter might be in some way related to such standing waves. While it was initially unclear why matter should be quantized as discrete particles, such as the electron, rather than inhabiting the continuum of possibilities available to the electromagnetic spectrum, its discrete nature suggested that only certain allowed standing-wave modes were resonant and therefore at least quasi-stable.

This chain of thought led to the basic resonance model and to the material covered in Chapters 1 and 2, where standing-wave resonances of the electromagnetic field are identified with various massive particles.

It was not clear why a particular standing-wave resonance should have the mass that it does. Why should the mass of the electron be 0.51 MeV? Progress was made by deferring this question and posing a related one: If the electron, which is the lightest known massive particle, were identified with the first-order (half-wavelength) standing-wave resonance, then what would be the nature of the next higher resonance? The answer, given in Chapter 1, is that the next allowed resonance would have three half-wavelength segments and a mass and size similar to the muon or pion.

Both the electron and the muon, having spin, are circularly polarized. It seemed possible to form a resonance similar to the muon but linearly polarized. This could be conceptually assured by combining half-muons of opposite spin. The vector sum of the spins would cause a planar (two-dimensional) electronic current to oscillate back and forth. This resonance was identified with the charged pion. Further, a neutral pion could be conceived of as half a positive and half a negative pion combined in quadrature. Conceptual sketches of the muon and pion were presented in Figure 1-2. From such simple representations one immediately appreciates why the Pauli exclusion principle should apply to spinning Fermi-Dirac particles, like the electron and muon, whose rotation sweeps out a definite volume, but not to spinless Bose-Einstein particles, like the pion, which are effectively two-dimensional and can stack one on top of another like a deck of cards, forming boson condensates at low temperatures or high pressures.

The next conceptual breakthrough occurred when it was realized that the phase velocity of the electromagnetic field circulating within the muon and pion resonances must be nine times greater than the speed of light. [See Equation (1-5b).]

This phase velocity augments the electromagnetic attraction between adjacent pions by a factor proportional to the square of the phase current, or $9^2 = 81$ times what one would expect of parallel electronic currents. This offers an explanation for the strong force, which has been experimentally estimated to be approximately two orders of magnitude stronger than standard electromagnetic forces.

Following simple geometric packing principles, one can devise composite structures for the proton, neutron, and other baryons. These structures consist of an outer ring of alternately charged pions—hexagonal in most cases, but possibly octagonal in the sigma and xi, and nonagonal in the omega—and a central cluster composed of a negative muon plus several bound pions. The size of such a structure would be similar to the size of the proton determined by scattering experiments. Significantly, it was shown in Chapter 2 that the masses of the baryons can be calculated to an accuracy of about one percent by using a simple semi-empirical formula. Furthermore, the magnetic moments of the baryons are easily estimated from these structures. The agreement with the observed moments of the proton and neutron is quite good. The moment of the as-yet unmeasured Σ^0 is predicted as a test of the model.

Another prospect raised by the resonance model is that the cold dark matter that appears to account for the majority of the mass of the universe may be associated with a heavy, stable, neutron-like composite particle formed, not of muons and pions, like the neutron, but of tau leptons and D_s mesons. Its rest mass is predicted to be approximately 9.2 GeV, and it should exhibit a detectable magnetic moment.

If one accepts the premise that both the muon and pion are purely electromagnetic resonances, then it is also possible to so characterize the mu neutrino, ν_μ, that is created when a charged pion decays,

$$\pi^+ \rightarrow \mu^+ + \nu_\mu \tag{1-9}.$$

Clearly, the neutrino can be seen to be the difference of two electromagnetic entities, so it too must be electromagnetic in origin. In fact, it was shown in Chapter 4 that neutrinos are traveling-wave modes of the electromagnetic field. They are quite similar to the photon in that they have zero rest mass and travel at the speed of light. However, neutrinos have spins of one half, versus the spin-one photon, and they have a much lower interactional cross-section with matter.

In order for a traveling electromagnetic wave to interact with matter, the initial and final reactants must match electromagnetic boundary conditions, considering all waves and particles from the same inertial reference frame. The pion in Equation (1-9) can decay into a somewhat larger muon because the muon is born with a linear momentum that, through Lorentz contraction, reduces it to the size of a pion in the center of momentum reference frame. This momentum has all the characteristics of a de Broglie matter wave.*

The neutrino must depart in the direction opposite the muon with a wavelength matching that of the original pion. Matching boundary conditions in this fashion is equivalent to conserving energy and linear momentum.

A significant issue resolved in Chapter 4 is the failure of parity conservation in pion decay. The resonance model requires that the directions of the linear momentum and axial spin in the muon should be oppositely directed, as observed. Parity conservation would require that the net helicity—the vector dot product of spin and linear momentum—be zero when averaged over a number of decays. (See Appendix D.)

Experiment establishes, as the resonance model would predict, that the emitted mu neutrino has a spin of ½ with negative (left-handed) helicity, so the recoiling muon gains a half unit of positive (right-handed) spin relative to the initial pion.

This parity violation suggested that the nature of the positive charge might be related to an extremely high-frequency right-handed pitch within the

---

* The relation $p = h/\lambda$ between linear momentum, p, in the form $|\frac{1}{2}L^+, \frac{1}{2}R^-\}$ from Equation (4-6c), and the de Broglie wavelength, λ, can be generalized rather easily. At rest, the standing-wave components of an electron have N half twists traveling to the right and an equal number traveling to the left. If we imagine that an electron has n additional half twists in the wave component going to the right, and n fewer half twists in the component going to the left, then the net twist (net charge) would be unaffected. The first n would modify the rightward component by $\{\frac{n}{N}R^-\}$, and the second would have the opposite effect of $\{\frac{n}{N}L^-|$ on the leftward component. From equation (4-5a), $\{\frac{n}{N}L^-| = |\frac{n}{N}L^+\}$. Thus the net effect would be the application of $|\frac{n}{N}L^+\} + |\frac{n}{N}R^-\} = |\frac{n}{N}L^+, \frac{n}{N}R^-\}$ to the electron, and the particle would appear to move toward the right with a velocity of $\frac{n}{N}c$. While we do not know that n is quantized, any steps in velocity would be minute, on the order of $(\Delta v)_{min} = c/N \approx 1.5 \times 10^{-11}$ cm/sec for $\Delta n = 1$. Detecting such quantization would prove very challenging.

A relationship between the de Broglie, Compton, and Planck wavelengths can be established from $\lambda = h/p = h/m_e v = (h/m_e c)(N/n) = (N/n)\lambda_C$. Using Equation (5-4), $\lambda_C = N\lambda_{Pl}$, we obtain $(\lambda)(\lambda_{Pl} n) = (\lambda_C)^2$. This is similar in form to the phase-velocity relationship $C'c' = c^2$, and it is tempting to associate λ and $n\lambda_{Pl}$ with phase-velocity and group-velocity characteristics of the electron, respectively.

electrodynamic motion of the standing wave. Similarly, a negative particle would have a high-frequency left-handed pitch. Such a possibility is suggested by the pioneering work of Kaluza and Klein discussed earlier. Their idea was that electronic charge could be the spillover of gravitation from a 'collapsed' fifth dimension. In the resonance model this idea was expanded into the third premise of this monograph, twist.

Each higher-order standing-wave resonant structure should be substantially smaller and more massive than the previous one. A chain of generations would eventually lead to a single primordial entity, suggested in Chapter 3 to have had some 4×10^{21} segments.

How could such a complex entity have originated? Perhaps it was a Fourier-like transform of the original void. As it decayed into lower-order resonant states, each new generation gained an internal unit of twist. It is the accumulated twist in the lowest generations that is realized as electronic charge and that can be equated with the existence of matter.

Electrons, at the very end of the cascade, would consist of a single segment with a negative charge manifest in the 2×10^{21} left-handed twists within its resonance structure. Muons would have one fewer twist, so their effective electronic charge should be one part in 2×10^{21} less than that of the electron.

The particles of the second-last generation were portrayed in Figures 1-2 as schematics of their localized resonant fields. The true extent of each particle and its associated electromagnetic field would fill the universe.

Electronic charge is believed to be an effect of the dynamic 'existence' currents in the twist of a particle. Particles with opposite charge create an attraction analogous to the force between wires carrying parallel currents. Their twists follow nearly parallel paths, analogous to the points of contact between two screws placed along side each other, when one has left-handed and the other right-handed thread. The threads have the same incline where they make line contact. The opposite would be true of two screws with same-handed thread, or of particles of similar electronic charge.*

---

* Taking this a step further, opposite spin orientations would reduce the repulsion between similarly charged particles. The fact that electron orbitals are each restricted to two electrons, which must have opposite spin, makes sense in this light, as orbitals are a ground state for the electrons in an atom.

The components of positronium have opposite charges, and so the effects should be reversed. Positronium exists in two ground states, corresponding to the two spin orientations of its components: antiparallel $^1S_0$ and parallel $^3S_1$. These have lifetimes of 8×10^{-9} and 7×10^{-6} seconds, and decay into 2γ and 3γ, repectively.[13]

While the electromagnetic forces between particles are determined by their internal left- or right-handed twist, and can be either attractive or repulsive, gravitational forces are determined solely by the numbers of segments, and so must always be positive. The gravitational attraction between electrons is a factor of 4×10^{42} less than the electromagnetic force. According to the resonance model, this ratio is dependent on the number of segments in the primordial resonance, and may be related to the mass of the universe.

A zero-sum universe is often described as a quantum fluctuation, similar to the emergence of a virtual particle from a vacuum. However, a fluctuation as massive as the universe should not have lasted as long as the universe already has. The description of creation presented here postulates that the primordial entity had zero mass, and therefore that the universe could have an indefinitely long lifetime.*

The transitory emergence of virtual electron-positron pairs is a dead end because these are the lowest possible order of particles. They cannot decay into anything else. They can do nothing but wait until their mutual attraction brings them back together and they return to the void. Virtual muon or pion pairs, being more massive, are less likely but even more transient. Their lifetimes are still long compared to the window of opportunity afforded by the uncertainty principle, and in general they too should re-annihilate before they have a chance to decay.

Presumably, then, the primordial entity split during its window of opportunity. Its daughters had slight masses and therefore more constrained existences than their parent. They too must have disintegrated before they could collapse back into the void from which they emerged. Evidently their velocities of separation were high enough to keep them apart this long.

Each subsequent decay would have added to the expanding sea of particles and radiation. The more decay products that were created, the less likely they were to all recoalesce and abort the universe. This accelerating entropy set the arrow of time.

The resonances of the first generation would have attracted one another with a gravitational force comparable in strength to the electrostatic force today, due to the number of segments in the first generation being equal to the number of twists in the electrons of the final generation. Conversely, the electrostatic force of the first generation would have been comparable in strength to the tiny gravitational force between electrons today, although repulsive because both daughters exhibited left-handed twist.

---

* The Heisenberg uncertainty principle states that $(\Delta E)(\Delta t) \geq \frac{1}{2}\hbar$. That is, a large mass energy implies a brief existence, at least for virtual particles.

In most speculations about the origin of a zero-sum universe, it is not clear where the mass of the universe is supposed to come from. According to the resonance model, mass is a function of the twist that developed during the decay of the primordial entity. It started out at zero, and grew with time.

Equations (5-3) and (5-4) can be combined to derive $e/N = G^{\frac{1}{2}}m_e$. The term e/N represents the smallest possible unit of electronic charge, the amount postulated to separate successive resonance generations (see Point 5 after Equation 3-8); whereas $G^{\frac{1}{2}}m_e$ may be thought of as the smallest unit of gravitational attraction. Assuming the particles of the first generation had approximately $2N = 4 \times 10^{21}$ segments, and that each segment contributed an equal $G^{\frac{1}{2}}m_e$ to their gravitational attraction, the masses of these particles would have been $Nm_e = 3.6 \times 10^{-6}$ gram. So, at the time that the primordial entity divided, the mass of the entire universe would have been about seven micrograms! As the branching continued, the numbers of decay products grew exponentially, eventually spawning the full mass of our universe.

Gravitation is the only force that could have reversed this expansion. Despite mass being poorly developed at the moment of creation, gravitation dominated the resonances. Electromagnetism was an incipient, rudimentary force. But this changed over time, as the growing twist reversed the ratio of gravitation to electromagnetism. Today, the residual gravitation is a ghost of what it once was.

5-IV. THEORY AND VERIFICATION

As the resonance model was fleshed out in the course of this work, an eye was kept out for ways to test it. A number of accessible predictions are summarized in the Predicted Results section of the Synopsis.

While one might think that empirical testing is a requirement of the scientific method, a rather unsettling post-modern trend in physics waives such reality checks in favor of mathematical consistency. This is highlighted by E. David Peat:

> [The] advanced theories of physics have little direct connection with anything that can be measured, and those experiments which are suggested by theory are probably decades away from being designed. Theories today are really emerging out of other theories, and their testing ground is no longer the experimentalist's laboratory but aesthetics, mathematical consistency, and their interrelationship to yet other theories. In addition, the mathematical language in which these theories are expressed has become so advanced that it is no longer always possible to give simple visual illustrations of what the theory means.

He offers a telling example:

> Yoichiro Nambu, the creator of the original string theory, has called this situation "Postmodern Physics." Theory has moved so far ahead of experiment that, he suggests, physics must now develop in new ways. When a new theory is created, rather than thinking in terms of crucial experiments and observations, physicists have to begin by investigating the theory's formal mathematical structure. The theory and its mathematical language are probed, recast, and related to other theories. Eventually it will be possible to discover its most fundamental form. So rather than seeking immediate contact with experiment, the idea is to advance the theory, expand its scope and make it consistent, and relate it to other theories.[14]

It is the authors' firm conviction that, while it is appropriate for researchers to delve into uncharted territory for considerable periods of time, in the end their work must pass empirical verification before being broadly accepted. Attempts to avoid these terms may serve an inner circle of specialists for a time, but are inadequate for the wider world. In this regard, string theories and the more inclusive M-brane theory should be viewed as quite tentative.

At the opposite extreme, the standard model has been crafted to match all experimental results, and it has weathered the rigors of testing surprisingly well over the past two decades. However, it is widely held that such an empirical model, with its many arbitrary parameters, will not evolve easily into a comprehensive theory.

The resonance model fits comfortably between these extremes.

5-V. CLOSING THOUGHTS

The ideas contained in this monograph evolved over the course of a decade. They were initially part of a vague attempt to describe electromagnetic fields without designating a reference frame. However, a point was soon reached when new experimental data and abstract concepts such as conservation of strangeness fit into the overall picture so well that they justified an increasing confidence in the model. The authors now have little doubt that troubling new data such as neutrino oscillations will eventually fit into place.

We may also expect the resonance model to lead ultimately to quantitative predictions for some of the basic constants of nature, such as c, $\hbar$, e, or m_e. However, much work remains to be done before such a point can be reached.

Final words do not come easily. However, an associate, Rich Newman, offered this quote,[15] which conveys the authors' feelings better than they could have expressed them:

> *I am satisfied with ... a glimpse of the marvelous structure of the existing world, ... striving to comprehend a portion.*

Albert Einstein, 1931

[1] Theodor Kaluza, 'On the unity problem of physics.'
Preußischen Akademie der Wissenshaften zu Berlin, **54** (1921), 966-972.
Cited in E. Witten, 'Search for a realistic Kaluza-Klein theory.'
Nuclear Physics B, **186**, 3 (10 August 1981), 412-428.

[2] O. Klein, 'Quantum theory and five-dimensional relativity theory.'
Zeitschrift für Physik, **37**, 12 (1926), 895.
Cited in Witten, *ibid.* Summarized in abstract 2541 of
Science Abstracts **29**, A, 11 (November 1926), 765-766.

[3] P.G. Bergmann, *Introduction to the Theory of Relativity*
(Prentice-Hall, New York, 1942), 268.

[4] *Ibid.,* 254-279.

[5] P.J.E. Peebles, *Principles of Physical Cosmology*
(Princeton University Press, 1993), 408;

P.C.W. Davies and J. Brown, *Superstrings: A Theory of Everything?*
(Cambridge University Press, 1988), 71-89;

S. Weinberg, 'A unified physics by 2050?'
Scientific American (December 1999), 68-75.

[6] Barry Parker, *Search for a Super Theory: From Atoms to Superstrings*
(Plenum Press, New York, 1987), 237-241.

[7] E. David Peat, *Superstrings and the Search for the Theory of Everything*
(Contemporary Books, New York, 1989), 55.

[8] Davies and Brown, 71-89.

[9] A.H. Guth, *The Inflationary Universe*
(Addison-Wesley, New York, 1997).

[10] B. Greene, *The Elegant Universe*
(W.W. Norton & Co., New York, 1999), 172-174;

Davies and Brown, 33-47.

[11] S. Weinberg, *ibid.*

[12] G.D. Coughlan and J.E. Dodd, *The Ideas of Particle Physics,* 2nd ed.
(Cambridge University Press, 1991), 187-188.

[13] R.B. Leighton, *Principles of Modern Physics*
(McGraw-Hill, New York, 1959), 264.

[14] Peat 275-276.

[15] Albert Einstein, *Ideas and Opinions*
(The Modern Library, New York, 1994).

Appendices

Appendix A

Mass Estimates of the Muon and Pion, and the Magnitude of the Strong Force

Due to the complexity of the fields that comprise the muon and pion resonances, no exact solutions are attempted in this appendix. Approximations are used to establish the rest masses and binding energies of these particles, and to demonstrate that electromagnetic factors are sufficient to account for the strong nuclear force.

The magnetostatic and electrostatic energies associated with an idealized spinning half-wavelength segment of charge $\frac{1}{3}e$ are determined first. This structure does not exist in itself, but can be considered to be a mathematical component of the muon and pion. Three such collinearly spinning segments form the muon. It will be seen that the combined stationary magnetic fields of the segments result in a substantial binding energy, which accounts for the mass difference between the muon and three non-interacting segments. The mass of such a segment was determined to be $243\ m_e = 124.2$ MeV in Chapter 1. A similar analysis is then presented for the pion. In this case the counter-rotating components of the pions cancel the axial magnetic field, so that their masses are significantly greater than the base value.

A SEGMENT OF CHARGE $\frac{1}{3}e$

The energy stored in a magnetic field, W_B, is

$$W_B = \frac{1}{2\mu_o} \int B^2 d\tau \tag{A-1,[1]}$$

where the magnetic field, B, is integrated over all differential volume elements, $d\tau$, and μ_o is the permeability of free space. (μ_o is an artifact of the MKS metric system, not a fundamental constant like e, $\hbar$, or c.)

To approximate the magnetic energy stored in one segment of a resonance, we assume that B is constant within a cylinder of radius a' and length $\lambda = 2\pi a'$, and zero elsewhere. Representing the cylinder with a solenoid of current I' and N' turns per meter,[2] where $N' = (\pi a')^{-1}$, and taking B to be the average value $\bar{B}$ within such a solenoid,

$$B \approx \bar{B} = \frac{\mu_0 I'}{\pi a'} \tag{A-2}.$$

Substituting into Equation (A-1) and integrating over the cylindrical volume $\pi^2(a')^3$,

$$W_B = \tfrac{1}{2}\mu_0(I')^2 a' \tag{A-3}$$

is the magnetostatic energy per segment. It may be assumed that

$$I' = \tfrac{1}{2\pi}e'\omega' \tag{A-4},$$

where ω' is the effective circulation frequency of the electronic charge e' in an presumed circular path of radius a'. In the footnote to page 6 it is shown that an electron is stable when its surface velocity is light speed. Its radius is therefore $a = \hbar/m_e c = 3.86\times10^{-13}$ m. This can be extended to the radius of a muon segment,

$$a' = \frac{\hbar}{m'c'} = \frac{\hbar}{81 m_e \cdot 9c} = \frac{a}{729} = 0.53\times10^{-15} \text{ m} \tag{A-5}.$$

The outside radius of a hexagonal ring of pions, such as the ones in the nucleonic structures proposed in Chapter 2, would be three times this, or 1.6×10^{-15} m, if we discount the binding effects between the segments. This is consistent with the radius of the proton, which has been determined from scattering experiments to be 1.2×10^{-15} m. [3]

Using Equation (A-5), the current I' is

$$I' = \tfrac{1}{2\pi}e'\omega' = \tfrac{1}{2\pi}e'(c/a)' = \tfrac{1}{2\pi}(e/3)(9c/a') = 4.33\times10^4 \text{ A} \tag{A-6}.$$

Substituting into Equation (A-3), and using the MKS relationship $c^2 = (\varepsilon_0 \mu_0)^{-1}$, the magnetic-field energy stored in a single isolated segment would be

$$W_B = \frac{3^4}{\pi}\left[\frac{(e')^2}{8\pi\varepsilon_0 a'}\right] = 3.89 \text{ MeV} \tag{A-7}.$$

The term in rectangular brackets is the electrostatic energy stored by a uniform charge of e' distributed over a spherical surface of radius a',

$$W_E = \frac{1}{8\pi\varepsilon_0} \frac{(e')^2}{a'} = +0.15 \text{ MeV} \qquad (A\text{-}8).[4]$$

The values of the magnetostatic and electrostatic energies are consequences of scaling the electronic charge to one-third of normal. These energies were implicitly included in Equation (1-4) in determining the $81\text{-}m_e$ mass of an idealized segment. However, interaction among the segments would create additional binding energies, as discussed below.

MU LEPTONS

In the case of the resonance identified with the muon in Chapter 1, the magnetic fields of the three half-wavelength segments are collinear, and their combined effect is thus the sum of their individual contributions. A precise estimate of the magnetic flux coupling between the segments is difficult. They would ideally be separated by a distance of $\pi a'$. However, from the perspective of the phase velocity, $C' = 9c$, the apparent spacing would be reduced to $\pi a'/9 \approx a'/3$. Thus the segments would appear to be so close together, relative to their diameters of $2a'$, that the combined magnetic field would approach $3\bar{B}$. Substituting this into Equation (A-1), the total magnetic energy stored in three idealized segments turns out to be

$$W_{3B} \approx 9W_B = 35.0 \text{ MeV} \qquad (A\text{-}9).$$

The difference between this and the magnetic energy of the three isolated segments,

$$3W_B - W_{3B} = -6W_B = -23.4 \text{ MeV} \qquad (A\text{-}10),$$

is the attractive magnetostatic binding energy generated by the parallel currents in the muon.

The electrostatic repulsion caused by bringing three charged segments together would be the sum of two electrostatic contributions, given by Equation (A-8), for the two pairs of adjacent segments, plus a reduced contribution for the effect of the end segments on each other. Making an initial guestimate for the reduced contribution of half that of the adjacent segments, the intersegmental repulsive electrostatic energy for the muon is estimated to be approximately 0.38 MeV.

This repulsive energy can be used, along with the magnetic binding energy determined above, to estimate the rest mass of the muon,

$$m_\mu \;=\; 243m_e \;-\; 6W_B \;+\; (2\tfrac{1}{2})W_E \;=\; 101.2 \text{ MeV} \qquad \text{(A-11)}.$$

$$ \underset{124.2}{} \quad \underset{23.4}{} \quad \underset{0.38}{\phantom{(2\tfrac{1}{2})W_E}}$$

This is off by about −4%. Given the number of approximations that were made, it appears that the binding energy estimates are of the proper magnitude, and that a spinning three-segment resonance is a good representation of the muon. This demonstrates that purely electromagnetic effects can account for the mass difference between the second-generation base value and the actual muon.*

PI MESONS

The spinless pi mesons are also identified as three-segment resonances in Chapter 2. These particles, formed from linearly polarized standing waves, may appear as shown in Figure 1-2(b). Due to their effectively counter-rotating charge components, the magnetic field is zero. This eliminates the magnetostatic binding energy of the muon, W_{3B}, from Equation (A-11). Equivalently, since the base mass of $243m_e$ assumes three W_B contributions for the three rotating segments, these need to be accounted for in the nonrotating pion. In the charged pions, the electrostatic W_E term remains:

$$m_{\pi^+} \;=\; 243m_e \;+\; 3W_B \;+\; (2\tfrac{1}{2})W_E \;=\; 136.2 \text{ MeV} \qquad \text{(A-12)}.$$

$$\phantom{m_{\pi^+} \;=\;} \underset{124.2}{} \quad \underset{11.7}{} \quad \underset{0.38}{\phantom{(2\tfrac{1}{2})W_E}}$$

This is reasonably close to the observed pion mass of 139.6 MeV. Here again it appears that electromagnetic effects are sufficient to explain the mass of a real particle.

---

* The electrostatic repulsion factor W_E may require adjustment according to the particle generation. Specifically, the $9c$ phase velocity within the muon and pion [see above and Equation (1-5b)] means that W_E may be nine times greater than in Equation (A-8), or +1.36 MeV. This brings the calculated mass of the muon to 104.2 MeV, within −1.4% of experiment. Furthermore, the relativistic contraction discussed in the text would result in a contribution closer to $3W_E$, bringing us to within just −0.7%.

The adjusted W_E increases the mass calculation of the charged pion to 139.3 MeV, off by only −0.2%, and accounts reasonably well for the mass difference between the charged and neutral pions: 3.4 MeV vs a measured 4.6 MeV. In light of these results, semi-empirical adjustment factors such as these may warrant further consideration.

THE STRONG NUCLEAR FORCE

The pion is the signature particle of the strong nuclear force. The attraction between pions in near planar contact constitutes much of the binding energy that holds nucleons together. Since leptons such as the muon are circularly polarized and can make only point contact with other particles, they do not participate in the strong force.

A similar attraction between pions in neighboring nucleons may be responsible for the residual nuclear force that binds atomic nuclei together. Indeed, in order to explain nuclear stability rules, nucleons can only bind 'with a limited number of [their] nearest neighbors'.[5] In other words, they must be in nearly direct contact for this residual force to operate, as one would expect from the mechanism proposed by the resonance model. This effect determines the relative numbers of protons and neutrons found in stable nuclei, since long-range electrostatic repulsion operates between *all* protons in the nucleus, not just those that are in contact.

When positive and negative pions approach each other, and their electronic currents are in phase, they will attract each other both electrostatically and electromagnetically. They will soon merge and transform into two linearly polarized electromagnetic waves. However, the mutual annihilation of a pion pair within the hexagonal ring of a proton is not energetically favorable. It would liberate only 279.2 MeV, less than the proton's binding energy of 284.3 MeV. The reaction is therefore endothermic and the proton is stable. The neutron, however, has an estimated binding energy of 279.2 MeV—almost identical to the mass energy of a pion pair—and is therefore just marginally unstable, with the exceptionally long lifetime of 890 seconds.*

Within composite particles, the electrostatic and in-phase electromagnetic attractions between pions form the binding energy. The electrostatic contribution per charged pion can be estimated by calculating the energy of three pairs of point charges spaced by $2a'$, or $3(-0.15 \text{ MeV}) = -0.45 \text{ MeV}$, from Equation (A-8). Although pions do not exhibit stationary magnetic fields, they should have high-frequency alternating fields associated with current flow transverse to the their axes. When averaged over time, the dynamic magnetic energy in each segment is estimated to be one-half of that given by Equation (A-7), or about -2 MeV. The combined fields of two

---

* This may explain not only why the neutron has such a long lifetime, but also why it is often stable within atomic nuclei. The binding energy in stable isotopes is evidently enough, at around 8 MeV per nucleon,[6] to stabilize the neutron against pion-pair decay. However, the exact mechanism of the neutron decay process $n \rightarrow p^+ + e^- + \bar{\nu}_e$, which involves the π^0 in the neutron core, is still unclear.

segments in adjacent particles would be approximately twice this. The magnetic binding energy, BE_B, for all three segment pairs of two adjacent pions would then be approximately

$$(BE_B)_\pi \approx (3)(2)(\tfrac{1}{2})W_B = -11.7 \text{ MeV} \tag{A-13},$$

and the total binding energy for two oppositely charged pions would be

$$BE_{\pi^*} = (BE_B)_\pi - 3W_E \approx -12.1 \text{ MeV} \tag{A-14}.*$$

The best fit to the observed rest masses of the composite subatomic particles listed in Table I of Chapter 2 was made by assuming neutral-pion binding energies of 12 MeV and charged-pion binding energies of 15 MeV. The results of Equations (A-13) and (A-14) are close enough to conclude that the strong force between pions may be entirely electromagnetic in origin.*

---

* With the relativistic adjustment to W_E considered in the footnote on page 103, the calculation of BE_{π^*} in Equation (A-14) increases to -15.8 MeV, quite close to our best fit of -15 MeV.

[1] P. Lorrain and D. Corson, *Electromagnetic Fields and Waves*, 2nd ed. (W.H. Freeman & Co., New York, 1962), 362.

[2] *Ibid.*, 298-299, 315-316.

[3] R.B. Leighton, *Principles of Modern Physics* (McGraw-Hill, New York, 1959), 549-551.

[4] Lorrain and Corson, 78.

[5] Leighton 551-561.

[6] G.D. Coughlan and J.E. Dodd, *The Ideas of Particle Physics*, 2nd ed. (Cambridge University Press, 1991), 42.

Appendix B

Segmental Annihilation Mechanisms in the Baryons

A negative muon and a coaxial positive pion are presumed to exist as electrostatically bound neighbors in the central clusters of all baryons. The more these particles overlap, the greater the electrostatic contribution to their binding energy. However, when full segments overlap, full destructive interference (mutual annihilation) could occur for an even greater effect on the binding energy.

At this point the overlap involves $\frac{1}{3}\mu^-$ and $\frac{1}{3}\pi^+$. Since the π^+ can be described as the sum of two components, $\frac{1}{2}\mu^{+\uparrow} + \frac{1}{2}\mu^{+\downarrow}$ (see Chapter 1), the overlapping segments may be rearranged as $\frac{1}{6}(\mu^{-\uparrow} + \mu^{+\uparrow})$ plus $\frac{1}{6}(\mu^{-\uparrow} + \mu^{+\downarrow})$. The first term could decay into oppositely directed and circularly polarized electromagnetic waves, and the second into linearly polarized waves. However, such waves could not couple to the ambient free-space electromagnetic field because their phase velocities are $9c$, according to Equation (1-5b). Nevertheless, they may oscillate as bound electromagnetic excitations along the axes of the baryons. If such an excited baryon were to collide with another particle, perhaps in the primordial meson soup, it would be able to transfer one half of its linear momentum. The recoil of the baryon would cancel the remaining linear excitation, but the spin would be unaffected.

When this happens, the remaining two segments of the π^+ would acquire the spin of the lost μ^- segment. The now-spinning pion would be unable to participate in any further annihilation with the muon, because to do so would reduce it to a single segment with twice the spin of a muon. (Muons already spin maximally, with surface current velocities of light speed, like the electrons discussed in Chapter 1.) However, a second pion could interact with either remaining muon segment. A second annihilation would form the core of a proton, if the second pion were another π^+, or the core of a neutron, if it were a π^0.

This mechanism may explain why baryons have only been observed being created in conjunction with antibaryons so that the total *baryon number* is conserved.[1] This is because such pair creation does not require any coupling to the free-space electromagnetic field or any collisional de-excitation.

The unstable conditions that existed during the initial stages of our universe may have favored the release of the energy from the lost segments of excited protons. It may have been released as kinetic energy during collisions between the meson clusters that apparently emerged from the dense 'meson sea' of the inflationary era. This possibility is discussed in Chapter 4, and may explain the observed preponderance of protons over antiprotons.

---

[1] G.D. Coughlan and J.E. Dodd, *The Ideas of Particle Physics*, 2nd ed. (Cambridge University Press, 1991), 51.

Appendix C

The Magnetic Moments of the Baryons

The resonance model provides sufficient detail that, in principle, the magnetic moments of composite particles may be calculated without ad hoc parameters. This contrasts with the standard model, which requires measurements of at least three baryons in order to calculate the moments of the others. The excellent agreement between observation and the results of the resonant model for the nucleons may provide insight into the remaining baryons.

MAGNETIC MOMENTS IN THE RESONANCE MODEL

If the muon presumed to exist in nucleon cores retained its free-space characteristics, the magnetic moments of these particles would be the same as the muon, since the pions that make up the rest of their structure do not contribute any spin. The magnetic moments of the nucleons would be that of a free muon, $\mu_\mu(m_p/m_\mu) = -8.88$ nuclear magnetons (μ_N). This clearly does not match the $+2.79$ μ_N of the proton or the -1.91 μ_N of the neutron.

The discrepancy may be resolved by considering the composite structures of the nucleons illustrated in Figure 2-6. In the case of the proton, only one third of the muon in the central cluster remains to contribute to magnetic effects, while the other two thirds of its angular momentum are divided between surviving segments of the two pions in the core. The magnetic moment of the proton can thus be roughly estimated by adding the coaxial contributions from the $\frac{1}{3}\mu^-$ and the two $\frac{2}{3}\pi^+$s as $-8.88(\frac{1}{3} - \frac{2}{3}) = +2.96$ μ_N. This estimate does not take into account the mass differences of the muon and pion contributions, but is nonetheless reasonably accurate.

Destructive interference, or segmental annihilation, is a more complicated matter in the neutron because one segment of the muon is consumed in interaction with a π^+, while a second segment is lost through interaction with a π^0. Assuming that the mechanism in the neutron is similar to the one in the proton, with the spin and charge of the second muon segment transferred to the two remaining segments of the π^0, then the magnetic moment of the neutron would be on the order of $-8.88(\frac{1}{3} - \frac{1}{3} + \frac{1}{6}) = -1.48\ \mu_N$. Again, the result is encouraging.

Since magnetic moments are highly sensitive to structural details, an attempt was made to refine these estimates by accounting more closely for these details.

The following assumptions were made for the proton:

A. The effective mass of the surviving muon segment equals the mass of the ideal charged segment given by Equation (1-4). That is, $m_{(\frac{1}{3})\mu} = 81\ m_e = 41.4$ MeV. This is the mass value expected when the attractive forces between adjacent segments of the muon are eliminated.

B. The effective mass of the two surviving π^+ segments is $m_{(\frac{2}{3})\pi^+} = (\frac{2}{3})m_{\pi^+} - 15$ MeV $= 78.0$ MeV. The binding energy per charged pion determined in the text, 15 MeV, is expected to be fully assigned to the two surviving segments of the π^+, because the lost segments extended well above the plane of the hexagonal pion ring, as illustrated in Figure 2-6, and therefore made minimal contributions to the proton's binding energy.

With these two assumptions, the magnetic moment of the proton can be calculated to be

$$\mu_p = \mu_\mu \left[\frac{1}{3}\left(\frac{1}{3}\frac{m_\mu}{m_{(\frac{1}{3})\mu}} \right) - \frac{2}{3}\left(\frac{2}{3}\frac{m_\mu}{m_{(\frac{2}{3})\pi^+}} \right) \right] = +2.82\ \mu_N \qquad \text{(C-1)},$$

where μ_μ is relative to the mass of the proton, $-(m_p/m_\mu)$. This falls within 1.1% of the observed moment of $+2.79\ \mu_N$.

The calculations for the neutron are based on similar assumptions. However, there are two possible ways of sharing the charge and spin of the lost muon segment with the two remaining segments of the π^0. Both cases are analyzed, with one providing a substantially closer fit to the observed magnetic moment of the neutron.

The following assumption is needed in addition to assumptions (A) and (B) used for the proton:

C. The contribution to the magnetic moment made by the π^0 has two possibilities:

Case I. The charge and spin are evenly distributed between the two surviving segments so that each segment has an effective charge of $-\frac{1}{6}e$. The mass of the two segments is $m_{(2/3)\pi^0} = (2/3)m_{\pi^0} - 12$ MeV $= 78.0$ MeV, with the binding energy of 12 MeV per π^0 developed in the text.

Case II. The charge and spin reside in one of the segments, presumably the one furthest from the muon segment due to electrostatic repulsion. The mass of this now-charged segment is assumed to be a full $(1/3)m_{\pi^-} = 46.5$ MeV because it is contiguous with the other π^0 segment but lies above the plane of the hexagonal pion ring. It is thus not expected to be significantly involved in the binding energy of the neutron.

The magnetic moment of the neutron can now be calculated to be

(Case I)

$$\mu_n = \mu_\mu \left[\frac{1}{3}\left(\frac{1}{3}\frac{m_\mu}{m_{(1/3)\mu}} \right) - \frac{1}{3}\left(\frac{2}{3}\frac{m_\mu}{m_{(2/3)\pi^+}} \right) + \frac{1}{3}\left(\frac{1}{3}\frac{m_\mu}{m_{(2/3)\pi^0}} \right) \right] = -1.19 \ \mu_N \quad \text{(C-2a)},$$

(Case II)

$$\mu_n = \mu_\mu \left[\frac{1}{3}\left(\frac{1}{3}\frac{m_\mu}{m_{(1/3)\mu}} \right) - \frac{1}{3}\left(\frac{2}{3}\frac{m_\mu}{m_{(2/3)\pi^+}} \right) + \frac{1}{3}\left(\frac{m_\mu}{m_{\pi^+}} \right) \right] = -2.09 \ \mu_N \quad \text{(C-2b)}.$$

Case II is the better match to the observed $-1.91 \ \mu_N$ moment of the neutron. The 9% discrepancy may be attributable to uncertainties in the various mass-value assumptions made, especially for the remote negatively charged segment of the core. It is also possible that the spin and charge from the canceled segment of the muon are not entirely confined to the one segment of the π^0, but spill over somewhat to the second.

On the other hand, the $-1.19 \ \mu_N$ of Case I is quite close to the magnetic moments observed for the Σ^- and Ξ^0 baryons, -1.16 and $-1.25 \ \mu_N$.[1] This may be attributed to the weaker coupling between the central components and the octagonal pion rings of the Σ^- and Ξ^0, which should be larger than the hexagonal ring of the neutron. Therefore Case I may more closely correspond to these baryons.

Due to an equivalent reduction in binding energy between the core and ring of the Σ^+, the effective masses of the surviving π^+ segments are expected to fall between their isolated values of 93.1 MeV and the 78.1 MeV used in calculating the moment of the proton. The observed $+2.46 \ \mu_N$ moment of the Σ^+ corresponds to an effective bonded pion mass, $m_{(\frac{2}{3})\pi^+}$ in Equation (C-1), of 83.9 MeV.

The magnetic moments of the Ξ^- and Λ can now be reasonably estimated by using this 83.9 MeV for the $m_{(\frac{2}{3})\pi^+}$ in Equation (C-2a), as well as by increasing the effective mass of the lower subassemblies shown in Figures 2-8 and 2-9 to $(\frac{5}{3})m_\pi$ to account for the extra pion. This results in a magnetic moment of -0.55 μ_N, off by -15% from the -0.65 μ_N measured for the Ξ^- and -9% from the -0.61 μ_N of the Λ. While these discrepancies cannot be ignored, they represent a substantial improvement over the results of the standard model.*

MAGNETIC MOMENTS IN THE STANDARD MODEL

The simplest version of the standard model assumes that the proton is made up of three point-like particles: two *up* quarks, u, each with charge $+\frac{2}{3}e$ and spin $\frac{1}{2}\hbar$; and one *down* quark, d, with charge $-\frac{1}{3}e$ and spin $\frac{1}{2}\hbar$. The up quarks are presumed to manifest a triplet spin state, χ ($J = 1$; $m = -1$, 0, $+1$), while the d-quark exists in a doublet state, ϕ ($J = \frac{1}{2}$; $m = -\frac{1}{2}$, $+\frac{1}{2}$).

The angular-momentum wave function of a spin-up proton is ψ ($J = \frac{1}{2}$; $m = +\frac{1}{2}$). Using standard quantum mechanical formalism for combining the χ and ϕ angular-momentum functions,

$$\psi(\tfrac{1}{2}, \tfrac{1}{2}) = (\tfrac{2}{3})^{\frac{1}{2}}\,\chi(1, 1)\,\phi(\tfrac{1}{2}, -\tfrac{1}{2}) - (\tfrac{1}{3})^{\frac{1}{2}}\,\chi(1, 0)\,\phi(\tfrac{1}{2}, \tfrac{1}{2}) \qquad (C-3).[2]$$

The magnetic moment of with the first term on the right-hand side of this equation corresponds to $\mu_u + \mu_u - \mu_d$; that of the second term is just μ_d. Hence the magnetic moment for the proton is

$$\mu_p = (\tfrac{2}{3})(2\mu_u - \mu_d) + (\tfrac{1}{3})\mu_d = (\tfrac{1}{3})(4\mu_u - \mu_d) \qquad (C-4a).$$

---

* There is very little data available for the spin-$\frac{3}{2}$ baryons, and the structures proposed in Chapter 2 are correspondingly tentative. Nonetheless, the PDG does report a magnetic moment of $+3.7$ to $+7.5$ μ_N for the Δ^{++}. (The only citation from the past decade, Lopez Castro & Mariano (2001), gives $+6.14 \pm 0.51$ μ_N.)[1] In the resonance model, the moment would be $(\mu_p + \mu_n + \mu_{\mu^+})\,m_{\Delta^{++}}/(m_p + m_n + m_{\mu^+})$, the sum of the moments of the spin components normalized for the mass of the baryon, and at $+6.06$ μ_N is right on the mark.

The only other spin-$\frac{3}{2}$ baryon with a reported magnetic moment is the unusually stable omega. If its spin constituents are indeed $\bar{p}$, n, n, as proposed, then its moment might be expected to be $(2\mu_n - \mu_p)\,m_\Omega/(2m_n + m_p) = -3.9$ μ_N. Instead, the PDG reports half this: -2.02 ± 0.05 μ_N. Bare in mind, however, that magnetic moments are highly sensitive to structural details. There may well be mediating factors in the omega, like those proposed for the sigma and xi baryons.

The magnetic moments of the other spin-½ baryons can be similarly derived using the quark compositions of Figure 2-4, where s denotes the *strange* quark:

$$\mu_p = (\tfrac{1}{3})(4\mu_u - \mu_d) \qquad\qquad \mu_n = (\tfrac{1}{3})(4\mu_d - \mu_u)$$

$$\mu_{\Sigma^+} = (\tfrac{1}{3})(4\mu_u - \mu_s) \qquad\qquad \mu_{\Sigma^-} = (\tfrac{1}{3})(4\mu_d - \mu_s)$$

$$\mu_{\Sigma^0} = (\tfrac{1}{3})(2\mu_u + 2\mu_d - \mu_s) \qquad\qquad \mu_\Lambda = \mu_s \qquad\qquad \text{(C-4b).}[3]$$

$$\mu_{\Xi^0} = (\tfrac{1}{3})(4\mu_s - \mu_u) \qquad\qquad \mu_{\Xi^-} = (\tfrac{1}{3})(4\mu_s - \mu_d)$$

In addition, the $\Sigma^0 \rightarrow \Lambda$ transition moment is

$$\mu_{(\Sigma^0 \rightarrow \Lambda)} = \tfrac{1}{\sqrt{3}}(\mu_d - \mu_u) \qquad\qquad \text{(C-4c).}[3]$$

At this point, the moments of three measured particles are needed to determine the moments of the three quarks. The Particle Data Group (PDG)[3] has chosen the proton, neutron, and lambda, and the results are

$$\mu_u = +1.852\ \mu_N$$

$$\mu_d = -0.972\ \mu_N$$

$$\mu_s = -0.613\ \mu_N.$$

These values are substituted into Equations (C-4) to calculate the magnetic moments of the remaining baryons. The results are in fair agreement with observation, but only to within about 10-20%. The greatest discrepancy is −24% for the moment of the Ξ^-, which is calculated to be −0.493 μ_N but observed to be −0.651 μ_N.

Over the years, a number of refinements have been attempted, for example by using relativistic wave functions. However, after a detailed review, D.W. Herzog *et al.* have concluded that all such attempts fall short of a complete understanding.[4]

Most of these calculations are *post hoc*. One is not: the magnetic moment of the Σ^0 has yet to be measured, and the standard and resonance models make contradictory predictions. According to the standard model, both the Σ^0 and Λ are composed of the same three quarks: an *up*, u; a *down*, d; and a *strange*, s; each having a spin of $\tfrac{1}{2}\hbar$. For the net $\tfrac{1}{2}\hbar$ spin of a baryon, two of the quarks must be spin up and a third spin down, with three permutations:

$$u^\uparrow, d^\downarrow, s^\uparrow \qquad\qquad u^\downarrow, d^\uparrow, s^\uparrow \qquad\qquad u^\uparrow, d^\uparrow, s^\downarrow$$

The first two permutations have *isotopic spins* (isospins) of zero since the u and d are oppositely directed and the s does not contribute to isospin. When combined as $[u^\uparrow, d^\downarrow, s^\uparrow - u^\downarrow, d^\uparrow, s^\uparrow]$, they form an antisymmetric isospin state that is associated with the Λ.[1] Therefore the u and d in the Λ make no contribution to the moment, and $\mu_\Lambda = \mu_s$ as given in Equation (C-4b). The third configuration has an isospin of one and is associated with the Σ^0.[1]

It is now evident why the change in magnetic moment during the transition $\Sigma^0 \rightarrow \Lambda$ is proportional to $|(\mu_d - \mu_u)|$: The parallel spins of the up and down quarks in the Σ^0 become antiparallel in the Λ, while the strange quark is unaffected. While the actual transition moment has not been measured, it's absolute value has. The PDG reports $|\mu_{(\Sigma^0 \rightarrow \Lambda)}| = 1.61 \pm 0.08 \ \mu_N$.[1]

The only way to fit this result into the assumptions of the standard model is to *assume* that the sign of the transition moment is negative, and this is how the data is generally reported.

This assumption is required because the PDG has independently made the assignments $+1.852 \ \mu_N$ for the up quark and $-0.972 \ \mu_N$ for the down. This leads to a calculated transition moment from Equation (C-4c) of $-1.63 \ \mu_N$. The end result is that the magnetic moment predicted by Equations (C-4b) for the Σ^0 is $+0.791 \ \mu_N$.

A weakness of the standard model is that it says nothing about how a simple redirection of quark spins could result in the substantial 77-MeV change in mass from the Σ^0 to the Λ. The resonance model, on the other hand, accounts for this with the segmental annihilation mechanism discussed in Chapter 2 and Appendix B. The structure proposed for the Σ^0 in Chapter 2 also implies a neutron-like magnetic moment, estimated later in this appendix to fall in the range -1.19 to $-2.09 \ \mu_N$.

Although contrary to the results of the standard model, a negative moment is in keeping with experimental observation. The magnetic moment of the $\Sigma^0 \rightarrow \Lambda$ transition can be represented in the resonance model by

$$\mu_{(\Sigma^0 \rightarrow \Lambda)} = \mu_\Lambda - \mu_{\Sigma^0} \qquad \text{(C-5)}.$$

The calculated range of the transition moment is -0.613 less -1.19 to -2.09, or $+0.58$ to $+1.48 \ \mu_N$. The higher of these numbers is near the lower end of the observed range, 1.6 standard deviations from the measured absolute value of the moment. Thus a simple measurement of the sign of the magnetic moment of the Σ^0 could serve as a major test of the validity of the standard model *vs* the resonance model. A negative μ_{Σ^0} would create a serious crack in the standard model that would point toward the need for new physics. On the other hand, a positive value would leave the standard model intact and force a rethinking of the structure proposed here for the Σ^0.

[1] K. Hagiwara *et al.*, 'The Review of Particle Physics.'
      Physical Review, **D66** (2002). Available online at *http://pdg.lbl.gov.*

[2] D.H. Perkins, *Introduction to High Energy Physics*, 2nd ed.
      (Addison-Wesley, Reading, MA, 1982), 201-203.

[3] C.G. Wohl, Baryon Magnetic Moments (1994).
      In the text of the Λ-particle data in Hagiwara *et al.*

[4] D.W. Herzog *et al.*, 'Exotic-atom measurement of the magnetic dipole moment
      of the Σ^- hyperon.' *Physical Review* D, **37**, 5 (1 March 1988), 1142-1152.

Appendix D

Parity Violation in Weak Interactions

The process of spatial inversion is called the *parity operation*. This is the change of spatial coordinates such as x, y, z into their opposites, $-x, -y, -z$, like a reflection in a mirror along the three orthogonal axes. When the basis of modern physics was developed in the first half of the twentieth century, all observed processes were invariant under parity operation. That is, they *conserved parity*. However, in 1957 various weak interactions involving neutrino emission were found to violate parity conservation.[1] In fact, parity violation has turned out to be "practically the *signature* of the weak force".[2] For example, observations of the decay of a π^+ into a μ^+ and a ν_μ [Equation (4-6c)] have shown that the recoil and spin directions of the μ^+ are antiparallel in the reference frame of the π^+.[3] That is, the recoiling μ^+ advances like a left-handed screw. Formally, this means that it has negative helicity, H, which is defined as the vector dot product of the spin, σ, and the unit momentum vector, $\hat{p} \equiv p/|p|$,

$$H \equiv \sigma \cdot \hat{p} \qquad \text{(D-1)}.$$

In order to conserve linear and angular momentum, the oppositely directed ν_μ must also have negative helicity.

If parity were conserved, this decay process would result in the μ^+ having positive or negative helicity with equal probability. This is because the average helicity indicates the degree to which the spin is aligned to the linear momentum. Put in mathematical terms, helicity is a pseudoscalar: it changes sign under parity operation (reflection) because p changes sign but σ does not. In order to conserve parity, the ensemble average $\langle H \rangle$ must therefore be zero.[4] Experimental evidence indicates that in the case of $\pi^+ \rightarrow \mu^+$ decay, parity is consistently violated: $\langle H \rangle$ is -1.

Equation (4-6c) predicts a negative helicity for the μ^+ decay product because the spin and recoil directions are antiparallel. Thus the associations made in that equation correctly characterize the observed parity violation.

Parity violation was disconcerting when first discovered, since Maxwell's equations of electromagnetism are generally believed to be invariant under parity operation.* Since an original motivation for the resonance model was to explain all forces in nature through electromagnetic principles, parity violation was a potential problem here as well. The dilemma was resolved by early work of Theodor Kaluza, who concluded in 1919 that Maxwell's equations can be derived from general relativity by adding to it a fifth 'dimension' or degree of freedom.[5] This fifth dimension is cylindrical and somehow 'collapsed' so as to not show up as a linear coordinate in the macroscopic world. Electromagnetism turns out to be a macroscopic spillover from this dimension.

To flesh out this concept, it was postulated that positive electronic charge may be due to a compact right-handed twist in this dimension, and negative electronic charge to an equally tight left-handed twist. This assumption is used in Chapter 4 to identify the four well known neutrinos. If this model is correct, a complete parity operation would involve reflections in all five dimensions. It would therefore require a reversal of electronic charge, since reflection in the collapsed dimension would change the chirality of twist associated with charge sign. Reversing of all five parameters is commonly referred to as CPT operation because it includes a conventional *parity* inversion, P, in the three spatial dimensions; a *temporal* reversal, T, in the fourth dimension; and a *charge* reversal, C, in the fifth dimension. So far as is known, all forces in nature are invariant under CPT operation.[6]

The tight helical nature of the collapsed fifth dimension may have originated in the decay of the primordial standing-wave resonance introduced in Chapter 3. It is estimated there that this entity had approximately 2×10^{21}

---

* Maxwell's equations are:

$$\nabla \cdot \mathbf{D} = \rho \qquad\qquad \nabla \cdot \mathbf{B} = 0$$

$$\nabla \times \mathbf{E} = -\frac{\partial \mathbf{B}}{\partial t} \qquad \nabla \times \mathbf{H} = \mathbf{J} + \frac{\partial \mathbf{D}}{\partial t}$$

The parity operator inverts the signs of spatial coordinates (x, y, z), and *polar* vectors, such as force, $\mathbf{F}$, and velocity, $\mathbf{v}$. However, it does not change the signs of *axial* vectors, such as torque, angular momentum, or magnetic field strength, which are the vector product of two polar vectors. For example, the magnetic field, $\mathbf{B}$, is determined by the Lorentz force on a moving charge, $\mathbf{F} = e(\mathbf{v} \times \mathbf{B})$. Since parity inverts both $\mathbf{F}$ and $\mathbf{v}$, it does not affect $\mathbf{B}$. Within Maxwell's equations, the parity operator inverts the signs of $x, y, z, \mathbf{J}, \mathbf{E}, \mathbf{D}, \nabla \times \mathbf{H}$, and $\nabla \cdot \mathbf{B}$; but not $\rho, t, \mathbf{B}, \nabla \times \mathbf{E}$ or $\nabla \cdot \mathbf{D}$; leaving the equations as a whole unchanged. However, this may not hold precisely for weak interactions, as here the magnitude of ρ may change by one part in $N = 2 \times 10^{21}$.

wavelengths. This magnitude is consistent with the estimated mass of the universe. It is also speculated in Chapter 3 that the primordial entity decayed in such a way that the original 2×10^{21} wavelengths turned into the internal twists of the cascade products. In other words, the electronic charges of low-order resonance particles such as the pion and electron are fossils of the 2×10^{21} wavelengths of the primordial resonance that began our universe.

If this is the case, then the spacing between the twists of the electron, which itself is the size of the Compton wavelength, $\lambda_C = \hbar/m_e c = 3.86 \times 10^{-11}$ cm, should be comparable to the Planck length, $\ell_{Pl} = (G\hbar/c^3)^{1/2} = 1.62 \times 10^{-33}$ cm, a key parameter in string and superstring theories.[7] Indeed,

$$\lambda_C = \frac{N \ell_{Pl}}{\alpha^{1/2}} \tag{D-2},$$

where α is the fine structure constant, $e^2/\hbar c$, and N is the number of twists in the electron. (See Chapter 5.) Thus a 'string' may be associated with one revolution of twist in the collapsed fifth dimension.

The multitude of other dimensions introduced by string theories is only necessary if the strong, weak, and gravitational forces have other than an electromagnetic origin. Clearly, then, the resonance model has much in common with string concepts. Yet, if valid, it would result in a considerable simplification of conventional string and superstring theories.

---

[1] C.S. Wu *et al.*, 'Experimental test of parity conservation in beta decay.'
     Physical Review, **105**, 4 (15 February 1957), 1413-15.

[2] D. Griffiths, *Introduction to Elementary Particles*
     (Harper and Row, New York, 1987), 123.

[3] P. Hartman, *Nuclear and High Energy Particle Physics*
     (Cornell University Press, Ithaca, NY, *c.* 1959-60), 90-92.

[4] P.C.W. Davies and J. Brown, *Superstrings: A Theory of Everything?*
     (Cambridge University Press, 1988), 33-47.

[5] G.D. Coughlan and J.E. Dodd, *The Ideas of Particle Physics*, 2nd ed.
     (Cambridge University Press, 1991), 191.

[6] *Ibid.*, 47-48.

[7] *Ibid.*, 178-179.

Appendix E

Decay Pathways of the Heavy Baryons

The binding energies calculated for the baryons are highly dependent on the chosen structural parameters. After their compositions had been worked out, it was noticed that the decay products of the transitory, or 'heavy', baryons are generally consistent with the binding energies predicted by the resonance model.

The binding energy, BE, of a baryon is calculated from its rest mass, m_b, by subtracting the masses of the various elementary pions and muons, $\Sigma m_{\pi\mu}$, that it is proposed to be composed of. These constituent mass values are:

Charged pion: $\quad m_{\pi^\pm} = 139.57$ MeV
Neutral pion: $\quad m_{\pi^0} = 134.98$ MeV
Muon: $\qquad\qquad m_{\mu^\pm} = 105.66$ MeV

The favorableness of a particular decay mode depends on the energy of the internal transformation being enough to lift these internal particles from the binding energy of the baryon to that of its composite decay products.

The binding energies of the light composite decay products are:

Kaon (K$^\pm$): $\quad$ BE $= 60.0$ MeV $\quad [m_{K^\pm} (493.68 \text{ MeV}) - 3m_{\pi^\pm} - m_{\pi^0}]$
Proton (p): $\quad$ BE $= 283.9$ MeV $\quad [m_p (938.27 \text{ MeV}) - 8m_{\pi^\pm} - m_{\mu^-}]$
Neutron (n): $\quad$ BE $= 278.1$ MeV $\quad [m_n (939.57 \text{ MeV}) - 7m_{\pi^\pm} - m_{\pi^0} - m_{\mu^-}]$

The heavy baryons* and their principal decay products are listed in Table I.

---

* Decay of the heavy baryons is mediated by the strong nuclear force, often with the emission of pions, the hallmark of strong decay. The lighter and stabler nucleons, on the other hand, transform into one another through the emission of neutrinos, the hallmark of weak decay. See Chapter 4 for the weak nuclear force.

TABLE I. *Economical decay modes of the heavy baryons.*[1]

All energies and 'masses' (mc²) are in MeV. The column of 'proposed components' lists the elementary particles proposed in the text to comprise the baryons. The 'binding energy', BE_b, is the difference between the mass of the baryon and these components. All decay branches known to occur with a frequency of at least 1% are listed, in addition to one less common pathway, (a). Note that the one spin-½ baryon with a lifetime briefer than the π^0, (b), does not require pion decay. The 'decrease in binding energy', ΔBE, is the difference between the BEs of the baryon and of its products. The 'net reaction' is the net transformation from the internal components of the baryon to those of its products.

Baryon	Mass $m_b c^2$	Proposed Components	Binding Energy $BE_b = m_b - \Sigma m_{\pi\mu}$	Decay Products	Frequency of Pathway (%)	Lifetime, τ (s)
Σ^+	1189.37	$10\pi^\pm\,1\mu^-$	312.0	$p^+\pi^0$ $n\,\pi^+$	51.6 48.3	$\}\,0.8\times10^{-10}$
Σ^0	1192.64	$9\pi^\pm\,1\pi^0\,1\mu^-$	304.1	$\Lambda\gamma$	100	(b) 7.4×10^{-20}
Σ^-	1197.45	$8\pi^\pm\,2\pi^0\,1\mu^-$	294.7	$n\,\pi^-$	99.8	1.5×10^{-10}
Λ	1115.68	$9\pi^\pm\,1\pi^0\,1\mu^-$	381.1	$p^+\pi^-$ $n\,\pi^0$	63.9 35.8	$\}\,2.6\times10^{-10}$
Ξ^0	1314.9	$9\pi^\pm\,2\pi^0\,1\mu^-$	316.8	$\Lambda\pi^0$ $\Sigma^0\gamma$ (a)	99.5 0.35	$\}\,2.9\times10^{-10}$
Ξ^-	1321.32	$8\pi^\pm\,3\pi^0\,1\mu^-$	305.8	$\Lambda\pi^-$	99.9	1.6×10^{-10}
Δ^{++}	1230.8	$9\pi^\pm\,1\pi^0\,3\mu^\pm$	477.3	$p^+\pi^+$	>99	5.9×10^{-24}
Δ^+	1231.5	$8\pi^\pm\,2\pi^0\,3\mu^\pm$	472.0	$p^+\pi^0$ $n\,\pi^+$	>99	5.9×10^{-24}
Δ^0	1233.4	$9\pi^\pm\,1\pi^0\,3\mu^-$	474.7	$n\,\pi^0$ $p^+\pi^-$	>99	5.6×10^{-24}
Δ^-	≈1232	$8\pi^\pm\,2\pi^0\,3\mu^-$	≈470	$n\,\pi^-$	>99	?
Σ^{*+}	1382.8	$10\pi^\pm\,2\pi^0\,3\mu^\pm$	599.8	$\Lambda\pi^+$ $\Sigma^+\pi^0$ $\Sigma^0\pi^+$	≈88 ≈12	$\}\,18\times10^{-24}$
Σ^{*0}	1383.7	$11\pi^\pm\,1\pi^0\,3\mu^-$	603.5	$\Lambda\pi^0$ $\Sigma^0\pi^0$ $\Sigma^+\pi^-$	≈88 ≈12	$\}\,18\times10^{-24}$
Σ^{*-}	1387.2	$10\pi^\pm\,2\pi^0\,3\mu^\pm$	595.4	$\Lambda\pi^-$ $\Sigma^0\pi^-$ $\Sigma^-\pi^0$	≈88 ≈12	$\}\,17\times10^{-24}$
Ξ^{*0}	1531.8	$11\pi^\pm\,3\pi^0\,3\mu^\pm$	725.4	$\Xi^0\pi^0$ $\Xi^-\pi^+$	≈100	72×10^{-24}
Ξ^{*-}	1535.0	$12\pi^\pm\,2\pi^0\,3\mu^-$	726.8	$\Xi^0\pi^-$ $\Xi^-\pi^0$	≈100	66×10^{-24}
Ω^-	1672.45	$10\pi^\pm\,5\pi^0\,3\mu^\pm$	715.1	ΛK^- $\Xi^0\pi^-$ $\Xi^-\pi^0$	67.8 23.6 8.6	$\}\,0.8\times10^{-10}$

The 'reaction energy', W_{Rx}, is the mass energy of this change. (Here, as elsewhere, photons are disregarded, as they do not affect the component tallies.) Finally, the 'excess reaction energy' is the amount of reaction energy beyond that needed for the change in binding energy, $W_{Rx} - \Delta BE$. For economy, this should be less than the energy of the smallest unit of decay, the disintegration of the 135-MeV π^0.

Decay modes which increase the binding energy (that is, have a negative ΔBE), such as (c), may occur spontaneously. In such cases the net reaction may be endothermic, as in (d). Several Δ and Ω branches, (e), are uneconomical. Others, numbered (1)-(4), involve a transformation of muons into pions. These are discussed on the following pages.

Decrease in Binding Energy $\Delta BE = BE_b - BE_{prod}$	Proposed Product Components	Net Reaction (disregarding γs)	Reaction Energy $W_{Rx} = (\Sigma m_{\pi\mu})_b - (\Sigma m_{\pi\mu})_{prod}$	Excess Rx Energy $W_{Rx} - \Delta BE$	Baryon
28 / 34	} $8\pi^{\pm}\ 1\pi^0\ 1\mu^-$	$\pi^+ \pi^- \to \pi^0$	144 {	116 / 110	Σ^+
(c) -77	(no change)	—	0	77	Σ^0
17	$8\pi^{\pm}\ 1\pi^0\ 1\mu^-$	$\pi^0 \to \emptyset$	135	118	Σ^-
97	$9\pi^{\pm}\ 1\mu^-$	$\pi^0 \to \emptyset$	135	38	Λ
103	$7\pi^{\pm}\ 2\pi^0\ 1\mu^-$	$\pi^+ \pi^- \to \pi^0$	144	41	
(c) -64	(no change)	—	0	64	Ξ^0
13	$9\pi^{\pm}\ 1\pi^0\ 1\mu^-$	$\pi^0 \to \emptyset$	135	122	
(c) -75	$10\pi^{\pm}\ 1\pi^0\ 1\mu^-$	$2\pi^0 \to \pi^+\pi^-$	(d) -9	66	Ξ^-
193	$9\pi^{\pm}\ 1\mu^-$	$\pi^0\,(\mu^+ \mu^-) \to \emptyset$	346	(e) 153	Δ^{++}
188 / 194	} $8\pi^{\pm}\ 1\pi^0\ 1\mu^-$	$\pi^0\,(\mu^+ \mu^-) \to \emptyset$	346 {	(e) 158 / (e) 152	Δ^+
197	$7\pi^{\pm}\ 2\pi^0\ 1\mu^-$	$2\pi^+\ 2\mu^- \to \pi^0$ (1)	355	(e) 159	Δ^0
191	$9\pi^{\pm}\ 1\mu^-$	$\pi^+ \pi^0\ 2\mu^- \to \pi^-$ (2)	346	(e) 156	
193	$8\pi^{\pm}\ 1\pi^0\ 1\mu^-$	$\pi^+ \pi^0\ 2\mu^- \to \pi^-$ (2)	346	(e) 153	Δ^-
219 / 288 / 296	} $10\pi^{\pm}\ 1\pi^0\ 1\mu^-$	$\pi^0\,(\mu^+ \mu^-) \to \emptyset$	346 {	128 / 58 / 51	Σ^{*+}
222 / 299	} $9\pi^{\pm}\ 2\pi^0\ 1\mu^-$	$2\pi^+\ 2\mu^- \to \pi^0$ (1)	355 {	133 / 56	Σ^{*0}
292	$11\pi^{\pm}\ 1\mu^-$	$\pi^+ \pi^0\ 2\mu^- \to \pi^-$ (2)	346	55	
214 / 291	} $10\pi^{\pm}\ 1\pi^0\ 1\mu^-$	$\pi^0\,(\mu^+ \mu^-) \to \emptyset$	346 {	132 / 55	Σ^{*-}
301	$8\pi^{\pm}\ 3\pi^0\ 1\mu^-$	$\pi^+ \pi^-\,(\mu^+ \mu^-) \to \pi^0$	355	55	
409 / 420	} $9\pi^{\pm}\ 3\pi^0\ 1\mu^-$	$(\pi^+ \pi^-)(\mu^+ \mu^-) \to \emptyset$	490 {	82 / 71	Ξ^{*0}
410	$10\pi^{\pm}\ 2\pi^0\ 1\mu^-$	$2\pi^+\ 2\mu^- \to \emptyset$ (3)	490	81	Ξ^{*-}
421	$8\pi^{\pm}\ 4\pi^0\ 1\mu^-$	$4\pi^{\pm}\ 2\mu^- \to 2\pi^0$ (4)	500	79	
274	$12\pi^{\pm}\ 2\pi^0\ 1\mu^-$	$3\pi^0\,(\mu^+ \mu^-) \to \pi^+\pi^-$	337	63	Ω^-
398	$10\pi^{\pm}\ 2\pi^0\ 1\mu^-$	$3\pi^0\,(\mu^+ \mu^-) \to \emptyset$	616	(e) 218	
409	$8\pi^{\pm}\ 4\pi^0\ 1\mu^-$	$\pi^0\,(\pi^+ \pi^-)(\mu^+ \mu^-) \to \emptyset$	625	(e) 216	

The decay products are largely consistent with the baryon structures proposed in the text, being just energetic enough to overcome the baryons' binding energies. The exceptions are the delta and omega baryons, which appear to convert more mass than needed. These are touched on below.

Notice that whenever a decay is favorable through a change in binding energy alone—that is, whenever the binding energies of the products are greater than those of the original particle—there is no net loss of constituent particles. In fact, there is no change of muons or pions at all in the cases of the neutral sigma and neutral xi. In the case of the negative xi, the process is apparently endothermic: The reaction $2\pi^0 \rightarrow \pi^+\pi^-$ actually absorbs $(\Sigma m_{\pi\mu}) = -9.2$ MeV of the energy released by the change in binding energy. The π^0 is equivalent to a superposition of a half π^+ and a half π^-, so $2\pi^0 \leftrightarrow \pi^+\pi^-$ could be expected to be a likely nondestructive transformation.

Unfavorable decays in terms of binding energy require some other source of energy to overcome the bonds of the parent baryon. In such cases either pair annihilation or π^0 disintegration may liberate the energy required. Where the energy needed to change the binding energy of the proposed structures is less than the 135-MeV rest mass of a π^0, the decay is economically triggered by the loss of a single π^0. This may occur as $\pi^0 \rightarrow \emptyset$ for 135 MeV, or in the form $\pi^+\pi^- \rightarrow \pi^0$ for to liberate 144 MeV. (Gamma rays have been disregarded in these schematic equations, as they do not alter the material constituents.) It is worth noting that the π^0 is its own antiparticle, the only particle with the option of radiative decay. That is, it is the only constituent that can spontaneously decay without interacting with another particle.

The spin-½ baryons generally have significantly longer lifetimes than the 0.84×10^{-16} second π^0. The one exception, the Σ^0, has no constituent change in its decay. That is, it does not depend on π^0 disintegration. These details are consistent with the baryon structures proposed in the text.

The spin-3/2 baryons are extremely unstable, being some of 'the most transient phenomena studied in the natural world'.*[2] They all decay into

---

* Except for the oddly stable Ω, the spin-3/2 baryons are so ephemeral that their traces cannot be captured in emulsions or cloud chambers in order to measure their lifetimes directly. Most references give their linewidth, Γ, instead. The lifetimes of the Δ, Σ^*, and Ξ^* in Table I were derived with the identity $\tau = \hbar/\Gamma$.[3]

Lorentzian functions describe both particle decay and oscillations within damped resonant circuits. Now, the number of oscillations in a circuit before attenuation is called its *Q-factor*. The linewidths of the delta baryons, $\Gamma = 111$ to 118 MeV, are about one tenth their 1230-MeV rest masses. This suggests a Q-factor of around 10. That is, there should be time for about ten oscillations within a delta before it flies apart.   *(continued)*

spin-½ products. Changing angular momentum from ³⁄₂ to ½ requires muon pair annihilation, as only the muons have spin. In the case of the delta baryons, ΔBE is less than $2m_\mu$, so muon annihilation alone should liberate enough energy for the decay. Yet there appears to be pion loss as well. (Consequently, the excess reaction energy, $W_{Rx} - \Delta$BE, is greater than m_π.)

Such inefficiencies were not predicted. But the deltas are the least stable of the baryons, barely hanging together long enough to be detected. They are also the only ones proposed to contain a full muon, unconstrained by destructive interference. (See Figure 2-10.) It may be that this free muon explains their instability. The inefficiency of the deltas may also be evidence of a two-step decay process: Perhaps π^0 disintegration is required to liberate a bound muon, and this must happen for it to annihilate the free muon.

The sigma-stars are similar to the deltas, but have an additional pion that binds a segment of the erstwhile free muon. They last three times as long as the deltas. In the xi-stars, a pion pins down a second segment of the muon; these last three times as long again. Such details may explain why instability does not correlate with mass in these baryons.

The Δ^0, Δ^-, and Σ^{*0} apparently involve a transformation of $2\mu^- \rightarrow 2\pi^-$. Recall from Chapter 1 and Equation (4-4) that a subcomponent equivalence of $2\mu^\pm$ and $2\pi^\pm$ was invoked for the nature of the pions, so this decay mechanism is consistent with the resonance model. Indeed, all decays in Table I are explicable in terms of pair annihilation and π^0 disintegration, when the equivalences of $2\pi^0 \sim \pi^+\pi^-$ and $2\mu^\pm \sim 2\pi^\pm$ are taken into account. The pathways involving the latter transformation are labeled (1) to (4) in the table and expanded on here. (As before, photons are disregarded.)

Notes to Table I.

A '~' signifies equivalent subcomponents, $2\mu^\pm \sim 2\pi^\pm$ or $2\pi^0 \sim \pi^+\pi^-$.

(1) $2\pi^+(2\mu^-) \sim 2\pi^+(2\pi^-) = (\pi^+\pi^-)(\pi^+\pi^-) \sim (\pi^+\pi^-)(\pi^0\pi^0) \rightarrow \pi^0$.

(2) $\pi^+\pi^0(2\mu^-) \sim \pi^+\pi^0(2\pi^-) = (\pi^+\pi^-)(\pi^0)(\pi^-) \rightarrow \pi^-$.

(3) $2\pi^+(2\mu^-) \sim 2\pi^+(2\pi^-) = (\pi^+\pi^-)(\pi^+\pi^-) \rightarrow \emptyset$.

(4) $3\pi^+\pi^-(2\mu^-) \sim 3\pi^+3\pi^- \sim (2\pi^+2\pi^-)(\pi^0\pi^0) \rightarrow 2\pi^0$.

But, assuming the deltas are close to the size of the proton, $r = 1.5 \times 10^{-13}$ cm (see Figure 2-10), the oscillation time would be $2r/c \approx 10^{-23}$ second, using up the full delta lifetime. The Q-factor should be closer to 1 than 10. That is, of course, if the maximum circulation velocity is c. If, however, the velocity is $9c$, as would be expected for a baryon (see Section 1-IV), then everything works out. This may provide additional support for the electromagnetic origin of the strong force.

Table I includes all decay branches with frequencies greater than about 1%. There are of course many less common branches. For example, whenever a particle generates a π^0 among its decay products, it may occasionally produce a gamma ray instead. This involves pair annihilation, $\pi^+\pi^- \rightarrow \emptyset$, and releases more energy than the economical $\pi^+\pi^- \rightarrow \pi^0$. But the π^0 is so extremely short lived compared to the charged pions and muons that we might expect there to be a slight chance of one disintegrating during the time it takes the other particles to merge and annihilate.*

The greatest difficulty is with the heaviest baryon, the omega. If the composition proposed in the text is correct, then the two less frequent xi modes are inefficient. They involve the destruction of one more constituent pion than energy conservation alone would require. But compared to the similarly inefficient deltas, the omega is quite stable, lasting as long as the spin-½ baryons.

The omega baryon has the most complex structure that has been proposed. It may be that this structure is not unique. But perhaps the omega's stability is related to the compact double ring illustrated in Figure 2-11, where there are just enough pions in the outer ring to encircle the core. This is reminiscent of the most stable radioactive nuclides, which have complete shells in their nuclei.

---

* These π^0-less branches are relatively uncommon: The Σ^+ produces $p^+\pi^0$ 51.6% of the time, but $p^+\gamma$ just 0.12% of the time. Likewise, the Λ produces 35.8% $n\pi^0$, but only 0.18% $n\gamma$; the Ξ^0 generates 99.5% $\Lambda\pi^0$, but only 0.10% $\Lambda\gamma$. These minor products may be explicable through a single mechanism. The only branch in Table 1 with less than a 1% frequency is the 0.35% $\Sigma^0\gamma$ product of the Ξ^0. This was included in the table because there is no corresponding $\Sigma^0\pi^0$ branch. However, since the Σ^0 and Λ baryons have the same internal components, $\Sigma^0\gamma$ may correspond to the $\Lambda\gamma$ mode of the Ξ^0.

The spin-½ baryons are similar: the deltas decompose into $N\pi$ (a nucleon plus pion) more than 99% of the time, but into $N\gamma$ just 0.5-0.6% of the time. The frequencies of the other baryons are not well known, but evidently the γ decay modes are uncommon.[1]

[1] Particle data are from K. Hagiwara *et al.*, 'The Review of Particle Physics.' *Physical Review*, **D66** (2002). Available online at *http://pdg.lbl.gov*.

[2] G.D. Coughlan and J.E. Dodd, *The Ideas of Particle Physics*, 2nd ed. (Cambridge University Press, 1991), 58.

[3] Gordon Kane. *Modern Elementary Particle Physics: The Fundamental Particles and Forces.* (Addison-Wesley, New York, 1993), 119-122.

Index

Index

Page numbers listed *in italics* refer to footnotes.
Page numbers **in bold** refer to figures or tables.

Index

About the Authors

Douglas Pinnow graduated with distinction from Cornell University as a Bachelor of Engineering Physics in 1961. While serving as a Naval Officer and Nuclear Engineer on Admiral Rickover's staff, he continued his studies in physics at The Catholic University of America in Washington, DC. These studies led to a PhD in physics as a NASA Fellow in 1967. Dr Pinnow went on to become a Member of the Technical Staff at Bell Telephone Laboratories in Murray Hill, New Jersey, where he served as Supervisor of the Quantum Electronics Group in the Solid State Device Development Laboratory.

After eight years at Bell Labs, Dr Pinnow became Assistant Manager of the Chemical-Physics Department at the Hughes Research Laboratory in Malibu, California. In 1979 he returned to the East Coast to become Director of R&D at Times Fiber Communications in Wallingford, Connecticut, and in 1983 he joined the Newport Corporation of Costa Mesa, California, as Vice President. Two years later he founded Universal Photonix in Laguna Hills as its president, and in 1995 founded and served as Chief Technical Officer of a subsidiary, Electronic Monitoring Systems. He has also served as an adjunct professor at UCLA and currently teaches a graduate course in electro-optics at UC Irvine.

Dr Pinnow has served as chairman for a number of major conferences sponsored by the Optical Society of America and the IEEE, including the Conference on Lasers and Electro-Optics (CLEO) in 1983. He has published more than fifty technical papers and has received numerous patents. In 1989 he was selected to become a Fellow of the Optical Society of America.

In his free time Dr Pinnow enjoys ocean sailing and racing. He has participated in the annual Newport-to-Ensenada race for the past ten years. He and his wife Joan serve as foster parents, often for premature babies with special problems such as prenatal drug exposure. During the years this monograph was in preparation, the Pinnows cared for over one hundred children.

Kirk Miller graduated as a Bachelor of Physics from UC Berkeley in 1990, then taught physics and chemistry for two years with the Peace Corps in the Republic of Benin (Dahomey). He received a Masters of Education at the University of Michigan, Ann Arbor, in 1994.

Besides hiking, shogi, and the tango, Kirk loves to travel. He lived in Japan for several years, and has spent time in Malaysia, Turkey, Slovakia, Korea, Nigeria, and Mexico. He is currently pursuing a Doctorate in Linguistics at UC Santa Barbara.

www.ingramcontent.com/pod-product-compliance
Lightning Source LLC
Chambersburg PA
CBHW032017170526
45157CB00002B/741